Lecture Notes in Control and Information Sciences

Edited by A. V. Balakrishnan and M. Thoma

Vol. 1: Distributed Parameter Systems: Modelling and Identification
Proceedings of the IFIP Working Conference, Rome, Italy, June 21–26, 1976
Edited by A. Ruberti
V, 458 pages. 1978

Vol. 2: New Trends in Systems Analysis
International Symposium, Versailles, December 13–17, 1976
Edited by A. Bensoussan and J. L. Lions
VII, 759 pages. 1977

Vol. 3: Differential Games and Applications
Proceedings of a Workshop, Enschede, Netherlands, March 16–25, 1977
Edited by P. Hagedorn, H. W. Knobloch, and G. J. Olsder
XII, 236 pages. 1977

Vol. 4: M. A. Crane, A. J. Lemoine
An Introduction to the Regenerative Method for Simulation Analysis
VII, 111 pages. 1977

Vol. 5: David J. Clements, Brian D. O. Anderson
Singular Optimal Control: The Linear Quadratic Problem
V, 93 pages. 1978

Vol. 6: Optimization Techniques
Proceedings of the 8th IFIP Conference on Optimization Techniques, Würzburg, September 5–9, 1977
Part 1
Edited by J. Stoer
XIII, 528 pages. 1978

Vol. 7: Optimization Techniques
Proceedings of the 8th IFIP Conference on Optimization Techniques, Würzburg, September 5–9, 1977
Part 2
Edited by J. Stoer
XIII, 512 pages. 1978

Vol. 8: R. F. Curtain, A. J. Pritchard
Infinite Dimensional Linear Systems Theory
VII, 298 pages. 1978

Vol. 9: Y. M. El-Fattah, C. Foulard
Learning Systems:
Decision, Simulation, and Control
VII, 119 pages. 1978

Vol. 10: J. M. Maciejowski
The Modelling of Systems with Small Observation Sets
VII, 241 pages. 1978

Vol. 11: Y. Sawaragi, T. Soeda, S. Omatu
Modelling, Estimation, and Their Applications for Distributed Parameter Systems
VI, 269 pages. 1978

Vol. 12: I. Postlethwaite, A. G. J. McFarlane
A Complex Variable Approach to the Analysis of Linear Multivariable Feedback Systems
IV, 177 pages. 1979

Vol. 13: E. D. Sontag
Polynomial Response Maps
VIII, 168 pages. 1979

Vol. 14: International Symposium on Systems Optimization and Analysis
Rocquentcourt, December 11–13, 1978;
IRIA LABORIA
Edited by A. Bensoussan and J. Lions
VIII, 332 pages. 1979

Vol. 15: Semi-Infinite Programming
Proceedings of a Workshop, Bad Honnef, August 30 – September 1, 1978
V, 180 pages. 1979

Vol. 16: Stochastic Control Theory and Stochastic Differential Systems
Proceedings of a Workshop of the „Sonderforschungsbereich 72 der Deutschen Forschungsgemeinschaft an der Universität Bonn"
which took place in January 1979 at Bad Honnef
VIII, 615 pages. 1979

Vol. 17: O. I. Franksen, P. Falster, F. J. Evans
Qualitative Aspects of Large Scale Systems
Developing Design Rules Using APL
XII, 119 pages. 1979

Vol. 18: Modelling and Optimization of Complex Systems
Proceedings of the IFIP-TC 7 Working Conference
Novosibirsk, USSR, 3–9 July, 1978
Edited by G. I. Marchuk
VI, 293 pages. 1979

Vol. 19: Global and Large Scale System Models
Proceedings of the Center for Advanced Studies (CAS) International Summer Seminar
Dubrovnik, Yugoslavia, August 21–26, 1978
Edited by B. Lazarević
VIII, 232 pages. 1979

Vol. 20: B. Egardt
Stability of Adaptive Controllers
V, 158 pages. 1979

Vol. 21: Martin B. Zarrop
Optimal Experiment Design for Dynamic System Identification
X, 197 pages. 1979

For further listing of published volumes please turn over to inside of back cover.

Lecture Notes in
Control and
Information Sciences
Edited by A.V. Balakrishnan and M.Thoma

57

T. Söderström
P.G. Stoica

Instrumental Variable Methods for System Identification

Springer-Verlag
Berlin Heidelberg GmbH 1983

Authors

Torsten Söderström
Dept. of Automatic Control and Systems Analysis
Inst. of Technology
Uppsala University
Uppsala, Sweden

Petre Gheorghe Stoica
Dept. of Automatic Control
Polytechnical Inst. of Bucharest
Bucharest, Romania

AMS Subject Classifications (1980): 93E12, 93B30, 90A16, 90A19, 62M10

ISBN 978-3-540-12814-4 ISBN 978-3-540-38743-5 (eBook)
DOI 10.1007/978-3-540-38743-5

Library of Congress Cataloging in Publication Data
Söderström, Torsten.
Instrumental variable methods for system identification.
(Lecture notes in control and information sciences ; 57)
Bibliography: p.
Includes index.
1. System identification. 2. Parameter estimation.
I. Stoica, P. G. (Petre Gheorghe)
II. Title. III. Series.
QA402.S693 1983 003 83-13525

Originally published by Springer-Verlag Berlin Heidelberg New York Tokyo in 1983

2061/3020-543210

FOREWORD AND ACKNOWLEDGEMENTS

This book is about system identification and parameter estimation using the instrumental variable (IV) approach. It has grown out of our joint research in this field during the last five years. Our attempt has been not only to summarize our findings in a unified form and extend them to some degree but also to include practical hints for the users. We have chosen a mathematical level of the text which we hope will appeal both to those interested in the technical details and those merely interested in how to apply IV methods in practice.

To understand the book the reader is assumed to have some basic knowledge of stochastic systems or time series analysis and also to know a little about parameter estimation in dynamic models. In places we assume the reader to be familiar with structures of multivariable systems. We have included various other background material in the appendices to make the book reasonably self-contained. For readers who are only practically interested we recommend a first reading of Part I and a careful reading of Part III, thus skipping the more theoretical Part II where the analysis is developed.

We have deliberately used rather few references in the text. Instead we have included some bibliographical notes in the end of the chapters. Since instrumental variable methods have been used for quite a long time and in diverse areas it should be needless to say that any attempt to give a full and perfect (historical) list of references is doomed to fail. We hope that we have succeeded in giving the key references and some more. Moreover, our purpose has been to present a unified analysis rather than tracing the exact historical background of every single variant of the instrumental variable approach.

We wish to express our sincere thanks to the many persons who helped make this book possible.

Several colleagues commented on the text. We would like to mention Professor Lennart Ljung in particular who made valuable suggestions based on a preliminary version of the manuscript. We also thank Professor Peter Young for his helpful comments on our work and for providing several references. Professor Norman Gough has suggested many improvements to the English. The responsibility for the remaining errors, English or technical, remains of course ours. We also thank

Professor Manfred Thoma, the editor of this series, who after preliminary discussions has encouraged us to write the book.

Most of our joint publications over the past years have been expertly typed by Miss Ingrid Ringård. We are very grateful to her and Mrs Lis Timner who jointly have typed the text so nicely with great care and patience. Mrs Timner has also skilfully prepared the figures.

We thank Dr Per Erik Modén who provided us with the drum drier data which we have used in Section 9.2.

The first author was given a scholarship by the Swedish Institute within an official exchange program between Sweden and Romania. This grant made possible a very fruitful stay in Bucharest. There we worked efficiently together on the spot thus avoiding the more cumbersome communication by mail.

Last, but not least we thank our families for their moral support during our work on this book.

TABLE OF CONTENTS

PART I: PRELIMINARIES AND DEFINITIONS

PART II: ANALYSIS

PART I

PRELIMINARIES AND DEFINITIONS

Chapter 1

INTRODUCTION

The area of "system identification", in particular "parameter estimation in dynamical models", is an important one in many fields, such as control engineering, econometrics and signal processing. There is certainly an overwhelming literature in this area. For an introduction to and a general survey of the area system identification we refer the reader to Åström and Eykhoff (1971), Box and Jenkins (1976), Eykhoff (1974), Goodwin and Payne (1977), Mehra and Lainiotis (1976). The present state of the art can be studied for example in Eykhoff (1981) and recent proceedings of the IFAC Symposia on Identification and System Parameter Estimation.

The area of identification is, as already indicated, related to other areas such as time series analysis, statistics and econometrics. A lot of various identification methods have been proposed and are in current use. A systematic way of comparison between various methods has been given to some degree but it is far from complete.

There seems to be an increasing interest in parametric methods, i.e. identification methods which as a crucial part include parameter estimation from experimental data. Two well-known parametric methods are the least-squares method (LSM) and the prediction error method (PEM). The LSM can be traced back to Gauss (1809). It is based on a simple algorithm. In fact, an analytic form of the estimate can be given once the data are known. A disadvantage of the LSM is that an asymptotic bias often occurs. This means that the estimated parameters often contain systematic errors that do not vanish even if the data series becomes infinitly long.

To avoid the asymptotic bias one possibility is to use a PEM in a model structure that also describes the properties of the noise. This leads to a more complex method and requires a numerical optimization of a nonlinear function that depends on the recorded data. The advantage of using such a method is that the estimated model often will give quite good a description of the data. In particular the bias obstacle is no longer a problem as for the simple LSM.

So far the discussion has indicated that the user must face a trade-off between algorithm complexity and the properties of the estimated parameters. It would of course be advantageous to have estimation methods that combine the small algorithmic complexity of the LSM and the good properties of the PEM estimates. The instrumental variable method (IVM) can be seen as an attempt to achieve this compromise. As we will see in this book the algorithmic complexity is quite low, while the estimates can have statistical properties that are superior to those of the LSM. For instance, the systematic errors of the parameter estimates go to zero as the data series length increases.

The IV method was apparently introduced by Reiersøl (1941). The method has been popular in the statistical field for quite a long period, cf Kendall and Stuart (1961). It has been applied to and adapted for dynamic systems both in econometrics and in control engineering. In the latter field pioneering work has been carried out by Wong and Polak (1967), Young (1965 a, 1970 a), Mayne (1967) and Rowe (1970). A well written historical background to the use of IVMs in econometrics and control can be found in Young (1976), who also discusses some refinement of the basic IVM.

The IV method as such covers many variants since the user has to specify several items within the method. The purpose of this book is to

- give a unified description of the various IV variants and to show how they are related
- analyse the (general) IV method with some common variants as special cases. The analysis will be concerned with asymptotic properties (convergence and asymptotic distribution of the parameter estimates)
- give hints for the user on how to apply IV methods in practice. For this purpose also some case studies will be included.

The three aims listed above correspond roughly to the three parts of the book.

The outline of the book is as follows:

In order to perform the analysis it will be most useful to have a pertinent framework. Thus in Chapter 2 we introduce and discuss the concepts of model structure *M*, system *S* and experimental condition *X*. These concepts are also discussed e.g. by Ljung (1976). In Chapter 2 we also introduce some basic assumptions that will be used throughout the text.

Chapter 3 is devoted to a description of various IV variants and a discussion of how they are related. Chapter 4 deals with consistency i.e. the limits of the parameter estimates when the number of data tends to infinity. The asymptotic distribution of the estimates is derived in Chapter 5. Optimization of the accuracy is discussed in Chapter 6. It is shown that by appropriately choosing the IV variant, an asymptotic accuracy comparable to that of the PEM can be obtained. Chapter 7 deals with optimal input design. It is shown that any possible covariance matrix of the parameter estimates can be obtained if the input is an autoregressive moving average (ARMA) process with appropriately chosen parameters. A practical algorithm based on this result is given as well.

In Part III, Chapters 8-9, we direct ourselves more to the practically oriented users. Thus in Chapter 8 we summarize and discuss the results of the analysis carried out in Part II. The reader who is not interested in *how* the results are derived should be able to read Part III directly after a first reading of Part I. Finally, in Chapter 9 some case studies are given.

In this text, equations, theorems and examples are numbered as a.b, where a refers to the chapter and b is a running index within the chapter.

Chapter 2

PRELIMINARIES

2.1 MODEL STRUCTURE

We start this chapter by discussing the model set or the *model structure M*.

A model will be described with a parameter vector θ. Several explicit examples of θ-vectors used to describe various models will be given later in this section. A model structure is a set of models, or (in an isomorphic way) a set of parameter vectors $\{\theta\}$.

Models can be of various kinds. In this book we will restrict the discussion to models that are linear in the parameters. This means that the model outputs are linear combinations of the unknown model parameters. We will formalize this requirement later as Assumption A1. This assumption is crucial for the instrumental variable approach. Note also that with a few exceptions we will only work with linear models, i.e. models where the output depends linearly on the past data. (The exception treated is the so called Hammerstein model, see Example 2.3). The transfer function of the linear model is to be estimated using a pertinent parameterization. In this book we will generally use black box models. For single-input single-output (SISO) models the parameterization is then almost trivial but for multivariable systems several possibilities exist. Most of the results in this book are applicable for multivariable processes. We will therefore make most of the assumptions and definitions applicable to the multivariable case. To facilitate the reading we have attempted to include also, whenever suitable, examples for SISO models.

For multivariable systems a most convenient parameterization of the transfer function matrix is the matrix fraction description, MFD. A general form of the model will then be

$$M\text{: } A(q^{-1},\theta)y(t) = B(q^{-1},\theta)u(t)+\varepsilon(t) \qquad (2.1)$$

where $y(t)$ is the output (an ny-dimensional vector) at time t (integer-valued), $u(t)$ is the input (an nu-dimensional vector), $\varepsilon(t)$ is the residual or equation error and θ is a $n\theta$-dimensional vector of unknown parameters. In (2.1) q^{-1} is the backward shift operator defined by $q^{-1}y(t) = y(t-1)$.

Further

$$A(z,\theta) = I+A_1(\theta)z+\ldots+A_{na}(\theta)z^{na}$$
$$B(z,\theta) = B_1(\theta)z+\ldots+B_{nb}(\theta)z^{nb} \quad (2.2)$$

where z is a (dummy) complex variable replacing the backward shift operator. Compatibility in (2.1) requires that $\{A_i(\theta)\}$, $A(z,\theta)$ are ny|ny dimensional matrices while the matrices $\{B_i(\theta)\}$, $B(z,\theta)$ have dimension ny|nu.

For convenience only no term $B_0(\theta)$ is included in (2.2). It is straightforward to extend the description as well as the analysis of this book to models with a $B_0(\theta)$ term. Similarly, models with a larger delay than one unit can also be used. If the delay is k (> 1) units then a new input $\tilde{u}(t) = u(t-k+1)$ can be defined by just shifting the input vector. The model

$$A(q^{-1},\theta)y(t) = B(q^{-1},\theta)\tilde{u}(t)+\varepsilon(t)$$

with $A(q^{-1},\theta)$ and $B(q^{-1},\theta)$ as in (2.2) can then be used. Also note that, as will be shown later, there is in a certain sense no restriction to take $A_0(\theta) = I$ as in (2.2).

The matrix coefficients $\{A_i(\theta)\}$ and $\{B_i(\theta)\}$ can depend on the parameter vector in various ways. We will give some examples below. We shall first, however, make the following general assumption on the model structure (2.1).

A1	The matrix coefficients $\{A_i(\theta)\}$, $\{B_i(\theta)\}$ are linear functions of θ.

Table 2.1 Assumption on the model structure M.

The assumption A1 is crucial for the application of instrumental variable methods. It is a consequence of A1 that the general model (2.1) can also be written as

$$M: y(t) = \phi^T(t)\theta+\varepsilon(t) \quad (2.3)$$

In (2.3) the matrix $\phi^T(t)$ is of dimension ny|nθ. Its entries are linear combinations of delayed input and output components.

We now complete this section with a few examples of model structures. These model structures will be referred to in the following.

Example 2.1 *A SISO model*

Assume that the model has one input and one output (nu = ny = 1). The model structure (2.1) is described by

$$M_1: A(q^{-1})y(t) = B(q^{-1})u(t)+\varepsilon(t) \tag{2.4a}$$

where

$$A(z) = A(z,\theta) = 1+a_1z+\ldots+a_{na}z^{na}$$
$$B(z) = B(z,\theta) = b_1z+\ldots+b_{nb}z^{nb} \tag{2.4b}$$

$$\theta = [a_1 \ldots a_{na}\ b_1 \ldots b_{nb}]^T \tag{2.4c}$$

Consequently

$$\phi^T(t) = [-y(t-1)\ldots -y(t-na)\ u(t-1)\ldots u(t-nb)] \tag{2.4d}$$

It is clear from (2.4d) how $\phi(t)$ depends on delayed input and output values. In this model structure we obviously let the parameter vector θ consist of all the coefficients $\{A_i\}$, $\{B_i\}$ in (2.2). ■

Remark on notation. When A_i, B_i are scalars they will in general be denoted by lower case letters: a_i, b_i. This convention has been used in (2.4b). Similarly when ny = 1 the matrix $\phi(t)$ becomes a column vector. We will then in general use the symbol $\varphi(t)$ for it. Also, we shall sometimes drop the argument θ in A_i, B_i for convenience. ■

Example 2.2 *A MISO model*

Assume that the model has one output (ny = 1). Then the model structure can be taken similar to that of Example 2.1. Since u(t) now is a column vector (of dimension nu) some care must be used. When the details are worked out we get

$$M_2: A(q^{-1})y(t) = B(q^{-1})u(t)+\varepsilon(t) \tag{2.5a}$$

$$A(z) = 1+a_1z+\ldots+a_{na}z^{na} \tag{2.5b}$$

$$B(z) = B_1 z + \ldots + B_{nb} z^{nb} \tag{2.5c}$$

$$\theta = [a_1 \ldots a_{na}\ B_1 \ldots B_{nb}]^T \tag{2.5d}$$

$$\phi^T(t) = [-y(t-1) \ldots -y(t-na)\ u^T(t-1) \ldots u^T(t-nb)] \tag{2.5e}$$

Note that here $B(z)$ and $\{B_i\}$ are row vectors of dimension nu. ■

Example 2.3 *A Hammerstein model*

As a special case of Example 2.2 we will consider a simple nonlinear model viz. the Hammerstein model. It will turn out that with appropriate redefinition this fits well into the framework of linear models used here. This model structure is given by

$$M_3:\ A(q^{-1})y(t) = B_1(q^{-1})\bar{u}(t) + B_2(q^{-1})\bar{u}^2(t) + \ldots + B_m(q^{-1})\bar{u}^m(t) + \varepsilon(t) \tag{2.6a}$$

where for convenience the input is denoted by $\bar{u}(t)$. The polynomials of M_3 are all scalar and given by

$$\begin{aligned} A(q^{-1}) &= 1 + a_1 q^{-1} + \ldots + a_{na} q^{-na} \\ B_i(q^{-1}) &= b_{i1} q^{-1} + \ldots + b_{inb} q^{-nb} \qquad i = 1, \ldots, m \end{aligned} \tag{2.6b}$$

Introducing a fictitious vector-valued input $u(t)$ and a row vector polynomial $B(q^{-1})$ through

$$u(t) = [\bar{u}(t)\ \bar{u}^2(t) \ldots \bar{u}^m(t)]^T \tag{2.6c}$$

$$B(z) = [B_1(z)\ B_2(z) \ldots B_m(z)] \tag{2.6d}$$

we get the model structure M_2, (2.5). ■

For multivariable i.e. multi-input multi-output (MIMO) models, the general model (2.1) can be parameterized in various ways. Most possibilities hinge on some special (canonical) forms of the matrix polynomial $A(z,\theta)$. Here we will discuss two simple model structures. Some other ones are mentioned in Section 2.5.

Example 2.4 Diagonal form for a MIMO model

This model structure is characterized by $A(z,\theta)$ being diagonal. It can be seen as a generalization of the model structure $\mathcal{M}_2$, (2.5), to the multivariable case. To be more specific we have

$$\mathcal{M}_4: \; A(q^{-1},\theta)y(t) = B(q^{-1},\theta)u(t)+\varepsilon(t) \tag{2.7a}$$

$$A(z,\theta) = \begin{bmatrix} a_1(z,\theta) & & \bigcirc \\ & \ddots & \\ \bigcirc & & a_{ny}(z,\theta) \end{bmatrix} \tag{2.7b}$$

where

$$a_i(z,\theta) = 1+a_{1,i}z+\ldots+a_{nai,i}z^{nai}$$

are scalar polynomials and

$$B(z,\theta) = \begin{bmatrix} B_1(z,\theta) \\ \vdots \\ B_{ny}(z,\theta) \end{bmatrix} \tag{2.7c}$$

where $B_i(z,\theta)$ is the following row-vector polynomial of dimension nu

$$B_i(z,\theta) = b_{1,i}z+\ldots+b_{nbi,i}z^{nbi}$$

Here the integer-valued parameters nai and nbi, i = 1,..., ny are the structural indices of the parameterization. The parameter vector θ and the matrix $\phi(t)$ can be written as

$$\theta = \begin{bmatrix} \theta_1 \\ \vdots \\ \theta_{ny} \end{bmatrix} \qquad \phi(t) = \begin{bmatrix} \varphi_1(t) & & \bigcirc \\ & \ddots & \\ \bigcirc & & \varphi_{ny}(t) \end{bmatrix} \tag{2.7d}$$

$$\theta_i = [a_{1,i} \ldots a_{nai,i} \; b_{1,i} \ldots b_{nbi,i}]^T$$

$$\varphi_i(t) = [-y_i(t-1)\ldots -y_i(t-nai) \; u^T(t-1)\ldots u^T(t-nbi)]^T$$

■

Example 2.5 ***Full polynomial form for a MIMO model***

This model structure

$$M_5:\ A(q^{-1},\theta)y(t) = B(q^{-1},\theta)u(t)+\varepsilon(t) \tag{2.8a}$$

is characterized by the fact that the parameter vector θ contains all the elements of $A_1(\theta),\dots, A_{na}(\theta), B_1(\theta),\dots, B_{nb}(\theta)$ in (2.2), viz.

$$\theta = \mathrm{col}\left(\begin{bmatrix} A_1^T \\ \vdots \\ A_{na}^T \\ B_1^T \\ \vdots \\ B_{nb}^T \end{bmatrix}\right) \tag{2.8b}$$

With col(A) we denote a column vector containing in order the columns of the matrix A. If A is a $m|n$ matrix then its ij:th element a_{ij} will appear as the $[m(j-1)+i]$:th element of col(A).

The matrix $\phi^T(t)$ becomes

$$\phi(t) = \begin{bmatrix} \varphi(t) & & \bigcirc \\ & \ddots & \\ \bigcirc & & \varphi(t) \end{bmatrix} = I\otimes\varphi(t) \tag{2.8c}$$

$$\varphi(t) = [-y^T(t-1)\dots\ -y^T(t-na)\ u^T(t-1)\dots\ u^T(t-nb)]^T \tag{2.8d}$$

In (2.8c) the symbol $\otimes$ denotes Kronecker product. If A is an $n_1|n_2$ matrix and B an $m_1|m_2$ matrix then $A\otimes B$ is the $(n_1m_1)|(n_2m_2)$ matrix given by

$$A\otimes B = \begin{bmatrix} a_{11}B & \dots & a_{1n_2}B \\ \vdots & & \vdots \\ a_{n_11}B & \dots & a_{n_1n_2}B \end{bmatrix}$$

A discussion of the Kronecker product can be found for example in Bellman (1970).

■

Note that for both the model structures M_4 and M_5 the matrix $\phi(t)$ is block diagonal. These models can therefore be considered as consisting of ny MISO models. (For M_5, though, the outputs enter formally as additional inputs). If these MISO models are denoted $M^{(1)},\dots, M^{(ny)}$ then $M^{(i)}$ and $M^{(j)}$, $i \neq j$ have no common parameters.

2.2 SYSTEM

In this section we will define and discuss *the system S*. With this concept we mean a mathematical description of the process to be identified. Such a description is of course an idealization. It should be stressed that to *define* and apply identification methods such as e.g. instrumental variable methods there is no need to assume a system description. However, in order to perform an *analysis* or to develop a *theory*, we need to introduce *pertinent* assumptions about the mechanism that generates the data. Our interest in the concept of "system" must be seen in this light.

Let us introduce some general assumptions on the system.

A2	The system S is linear, finite order, asymptotically stable and stochastic. The output can be written as S: $y(t) = x(t)+w(t)$ $x(t) = G(q^{-1})u(t)$ (2.9) where $x(t)$ is the noise-free output, $w(t)$ a disturbance and $G(q^{-1})$ the transfer function matrix.
A3	The disturbance $w(t)$ in (2.9) is a stationary stochastic process with zero mean and a rational, nonsingular spectral density matrix.

Table 2.2 Assumptions on the system S.

In A3 it would be sufficient to assume that the spectral density matrix is nonsingular for almost all frequencies ω. The difference in practice from the above formulation is, however, negligible. Note that $w(t)$ with a nonsingular spectral density matrix is sometimes referred to as a full rank process.

We then proceed to give some other general assumptions that concern both the system S and the model structure M.

The assumptions in Table 2.3 can be interpreted in some different ways.

A4	There exists a vector θ^* such that $A(z,\theta^*)^{-1}B(z,\theta^*) \equiv G(z)$ ($A(z,\theta)$, $B(z,\theta)$ being introduced in (2.1) and $G(z)$ in (2.9)).
A5	θ^* in A4 is unique

Table 2.3 Assumptions on the system *S* and the model structure *M*.

When the *model structure* M is regarded as given, Assumptions A4 and A5 give conditions on the system *S*. Assumption A4 means simply that there is a vector θ^* of "true values" that gives a perfect description of the deterministic part of the system, i.e. of the transfer function $G(\cdot)$.

When the *system* is regarded as given and fulfilling A2, then Assumptions A4 and A5 become conditions on the model structure. If na and nb in (2.1), (2.2) are sufficiently large Assumption A4 can always be fulfilled. (Remember that in A2 we assumed the true system *S* to be of finite order). Assumption A5 means in practice that the model has not been overparameterized (basically that na and nb have not been chosen too high). A strict analysis of A5 is rather more complicated and we will return to it later in this section for some common types of model structure. Also note that it is not restrictive to assume $A(0,\theta) = I$ in (2.2), in the sense that this constraint as such does not prevent any $G(z)$ from fulfilling A4.

It is now a consequence of Assumption A4 that the system (2.9) can be rewritten as

$$S: \quad \begin{aligned} y(t) &= \phi^T(t)\theta^* + v(t) \\ v(t) &= A(q^{-1},\theta^*)w(t) \end{aligned} \qquad (2.10)$$

cf the general model structure (2.1).

It is a consequence of Assumption A3 and the spectral factorization theorem, see e.g. Åström (1970), Anderson and Moore (1979), that the disturbance can be described as

$$v(t) = H(q^{-1})e(t) \qquad (2.11)$$

where $H(q^{-1})$ is a uniquely defined $ny|ny$ dimensional filter with $H(0) = I$ and with the property that both $H(q^{-1})$ and $H(q^{-1})^{-1}$ are asymptotically stable, and $e(t)$ is zero mean white noise (a sequence of independent and identically distributed random vectors) so that

$$E[e(t)e^T(s)] = \Lambda\delta_{t,s} \tag{2.12}$$

In (2.12), Λ is the covariance matrix of e(t) and $\delta_{t,s}$ is the Kronecker delta. As w(t) is a full rank process it follows that Λ is positive definite.

We will now consider the relation between the system *S* and the model structure *M*, as expressed with Assumptions A4-A5. We do this with some examples that parallel the examples in Section 2.1.

Example 2.6 Assumptions for the SISO case (Example 2.1 cont'd)

Assume that the transfer function G(·) (of the true system) can be written as

$$G(z) = \frac{B^*(z)}{A^*(z)} \tag{2.13a}$$

where

$$\begin{aligned} A^*(z) &= 1+a_1^*z+\ldots+a_{na^*}^*z^{na^*} \\ B^*(z) &= b_1^*z+\ldots+b_{nb^*}^*z^{nb^*} \end{aligned} \tag{2.13b}$$

are coprime polynomials. The identity in Assumption A4 then becomes

$$\frac{B(z,\theta)}{A(z,\theta)} \equiv \frac{B^*(z)}{A^*(z)} \tag{2.14}$$

Since $A^*(z)$ and $B^*(z)$ are coprime, A4 apparently means

$$n^* \triangleq \min(na-na^*,nb-nb^*) \geq 0 \tag{2.15a}$$

Moreover, A5 means

$$n^* = 0 \tag{2.15b}$$

The reason is that if $n^* > 0$, then $na > na^*$, $nb > nb^*$ and there must be a cancelling factor in the left hand side of (2.14). In fact if $n^* > 0$ the general solution of (2.14) with respect to θ for given $A^*(z)$, $B^*(z)$ is given by

$$A(z,\theta) \equiv A^*(z)L(z) \qquad B(z,\theta) \equiv B^*(z)L(z)$$

$$L(z) = 1+\ell_1 z+\ldots+\ell_{n^*} z^{n^*} \tag{2.16}$$

where L(z) has all zeros strictly outside the unit circle (due to A2) but is else arbitrary. Clearly, θ^* is then not unique. When (2.15b) is satisfied, θ^* is unique and given by

$$\theta^* = [\underbrace{a_1^*\ldots\ a_{na^*}^*\ 0\ldots\ 0}_{\text{na elements}} \mid \underbrace{b_1^*\ldots\ b_{nb^*}^*\ 0\ldots\ 0}_{\text{nb elements}}]^T$$

Note that zeros can appear in the first or the second half of θ^* but not in both. ■

Remark. The MISO model introduced in Example 2.2 will give exactly the same results as in Example 2.6 even if now $B^*(z)$ and $B(z,\theta)$ are row vector polynomials. The same results apply also to the diagonal form for MIMO models introduced in Example 2.4. We then simply have to interpret the model as a set of ny MISO models. Note that any finite order multivariable transfer function G(z) can be written in the form

$$G(z) = \begin{bmatrix} B_1^*(z)/a_1^*(z) \\ \vdots \\ B_{ny}^*(z)/a_{ny}^*(z) \end{bmatrix}$$

with $a_i^*(z)$, $B_i^*(z)$ being coprime. Here $a_i^*(z)$ is simply the least common denominator of the entries in the i:th row of G(z). The order relation (2.15a) now becomes

$$n_i^* \triangleq \min\ (nai-nai^*,\ nbi-nbi^*) \geq 0 \qquad i = 1,\ldots,\ ny \tag{2.17}$$

with $nai^* = \deg a_i^*(z)$, $nbi^* = \deg B_i^*(z)$. For A5 to hold we must have equality in (2.17) for all i. ■

We now turn to the full polynomial form for MIMO systems introduced in Example 2.5. Since this is a pseudocanonical rather than a canonical form its analysis becomes more complex. In the analysis we will use some concepts pertaining to rational matrices described in Appendix 6. We have the following result.

Lemma 2.1 Let the transfer function $G(z^{-1})$ be strictly proper (i.e. $G(0) = 0$). Then the following statements are equivalent:

i) The full polynomial form (2.8) fulfils Assumption A5 (θ^* exists and is unique)

ii) For given na, nb let $G^*(z) = z^{nb-na}G(z^{-1})$. The observability index ν[1] and the minimal degree δ[1] of $G^*(z)$ fulfil

$$\nu = na \tag{2.18a}$$

$$\delta = \nu \cdot ny \tag{2.18b}$$

iii) There exist polynomial matrices $A(z)$, $B(z)$ of degree na respectively nb such that

$$G(z) = A^{-1}(z)B(z) \tag{2.19a}$$

$$A(z),\ B(z) \text{ left coprime} \tag{2.19b}$$

$$\text{rank } [A_{na}\ \ B_{nb}] = ny \tag{2.19c}$$

Proof. We shall prove that i) ⇔ ii) and ii) ⇔ iii). Consider first the equivalence i) ⇔ ii). Introduce the following polynomial matrices

$$X(z) = X_1z+\ldots+X_{na}z^{na} \qquad Y(z) = Y_1z+\ldots+Y_{nb}z^{nb}$$

of dimensions ny|ny respectively ny|nu. It can easily be shown that i) holds if and only if

$$X(z)G^*(z) \equiv Y(z) \tag{2.20}$$

implies $X(z) = 0$, $Y(z) = 0$, with $G^*(z)$ defined above. Let $A(z)^{-1}\mathcal{B}(z)$ be the row proper factorization[1] of $G^*(z)$ with

$$A(z) = A_0+A_1z+\ldots+A_\nu z^\nu$$

$$\mathcal{B}(z) = \mathcal{B}_0+\mathcal{B}_1z+\ldots+\mathcal{B}_{\nu+nb-na-1}z^{\nu+nb-na-1}$$

Note that $\mathcal{B}_k = 0$, $k \geq \nu+nb-na$, follows from our assumption that $G(z^{-1})$ is strictly proper which here means that $\lim_{z\to\infty} z^{na-nb}G^*(z) = 0$

---

1) See Appendix 6 for definition and discussion.

Since $A(z)$ and $B(z)$ are left coprime, there exist polynomials $M(z)$ and $N(z)$ such that

$$A(z)M(z)+B(z)N(z) \equiv I$$

cf e.g. Kailath (1980). It then follows that, cf (2.20)

$$X(z)M(z)+Y(z)N(z) \equiv X(z)A(z)^{-1}$$

which shows that the solutions of (2.20) are given by

$$\begin{aligned} X(z) &\equiv L(z)A(z) \\ Y(z) &\equiv L(z)B(z) \end{aligned} \tag{2.21}$$

where $L(z)$ is a polynomial matrix restricted only by the condition that the degrees of the right hand sides of (2.21) are na and nb, respectively, and that $L(0) = 0$.

Assume first that (2.18) holds. Then since A_ν is nonsingular it easily follows that $L(z) \equiv 0$ and thus ii) $\Rightarrow$ i). Assume next that $na > \nu$. Then, regardless the value of δ we can take $L(z)$ as an arbitrary polynomial of degree $na-\nu$. Thus i) $\Rightarrow$ $na \leq \nu$. Moreover, na cannot be less than ν. Otherwise $[z^{na}A(z^{-1})]^{-1}[z^{nb}B(z^{-1})]$ would be a left factorization of $G^*(z)$ with the degree of the denominator $na < \nu$ and this is clearly impossible. We thus must have $na = \nu$. Finally assume that $na = \nu$ but $\delta < \nu \cdot ny$. Then A_ν is nonsingular. Let $L_1 \neq 0$ be a $ny|ny$-matrix, the rows of which belong to the null space of A_ν^T. Take

$$X(z) = L_1 zA(z) = L_1 A_{\nu-1} z^\nu + \ldots + L_1 A_0 z$$

$$Y(z) = L_1 zB(z) = L_1 B_{nb-1} z^{nb} + \ldots + L_1 B_0 z$$

Since $A(z)$ and $B(z)$ are left coprime we necessarily have rank $[A_0 \; B_0] = ny$ and this implies that $X(z)$ and $Y(z)$ above form a nontrivial solution of (2.20). Thus, i) $\Rightarrow$ (2.18); and the proof of equivalence i) $\Leftrightarrow$ ii) is complete.

Consider next the second equivalence ii) $\Leftrightarrow$ iii). Assume that iii) holds. It then follows from Lemma A2.1 given in Appendix 2 that the polynomial matrices

$$A^*(z) = z^{na}A(z^{-1}) = Iz^{na} + A_1 z^{na-1} + \ldots + A_{na}$$

$$B^*(z) = z^{nb}B(z^{-1}) = B_1 z^{nb-1} + \ldots + B_{nb}$$

are left coprime. As a consequence $A^*(z)^{-1}B^*(z)$ is the (normalized) row proper factorization of $G^*(z)$ and then (2.18) follows.

The implication ii) → iii) follows similarly by applying Lemma A2.1 to the normalized (i.e. with $A_\nu = I$) row proper factorization of $G^*(z)$. ■

Hence, in the case of the full polynomial form, Assumption A5 is not only a condition on the model orders. It may be inferred for example from (2.18) that A5 cannot be fulfilled for all transfer function matrices $G(\cdot)$. Fortunately it can be shown, Hannan (1976), that A5 can be satisfied generically (i.e. for almost all $G(\cdot)$ transfer function matrices). This can in fact be easily seen from (2.19). Finally, note that some further comments on the above result and the full polynomial form can be found in Section 2.5.

2.3 EXPERIMENTAL CONDITION

In this section we will discuss *the experimental condition X*. This concept is used to describe under what condition the identification experiment is carried out. In particular the input signal properties must be specified. For that purpose we define the concept of persistently exciting signals.

Definition 2.1 A signal $u(t)$ is said to be persistently exciting (p.e.) of order n if the limits

$$m = \lim_{N\to\infty} \frac{1}{N}\sum_{t=1}^{N} u(t) \tag{2.22a}$$

and

$$r_u(\tau) = \lim_{N\to\infty} \frac{1}{N}\sum_{t=1}^{N} [u(t)-m][u(t+\tau)-m]^T \qquad \text{(all } \tau\text{)} \tag{2.22b}$$

exist with probability one, and the matrix

$$R_n = \begin{bmatrix} r_u(0) & \dots & r_u^T(n-1) \\ \vdots & & \vdots \\ r_u(n-1) & \dots & r_u(0) \end{bmatrix} \tag{2.23}$$

is positive definite. ■

Note that for convenience we require (2.22b) to hold for all τ, even though only

$r_u(0),\ldots, r_u(n-1)$ appear in (2.23). This makes our future analysis simpler. In practice, (2.22) is a quite reasonable assumption.

For a stationary stochastic process with a rational spectral density, m will simply be the mean value and $r_u(\tau)$ the covariance function, i.e.

$$m = Eu(t) \tag{2.24a}$$

$$r_u(\tau) = E[u(t)-m][u(t+\tau)-m]^T \tag{2.24b}$$

see Söderström (1975).

The limits in (2.22) exist also for a deterministic and periodic signal. In that case the symbol "E" in (2.24) should be interpreted as $\frac{1}{M}\sum_{t=1}^{M}$ where M is the period of the signal, cf, e.g., Söderström (1975). Note that a process u(t) fulfilling (2.24) is called *ergodic* with respect to its second order moments.

In Appendix A1 we have given several useful results on persistently exciting signals.

We now present the general assumptions on the input signal. These assumptions are listed in Table 2.4.

A6	The input is a stationary process that is ergodic with respect to second order moments and persistently exciting.
A7	The input u(t) and the disturbance v(s) are independent for all t and s.

Table 2.4 Assumptions on the experimental condition $\mathcal{X}$.

Note that we have not specified within A6 of what order u(t) must be persistently exciting. The reason is that the precise requirement on u(t) will depend on the model structure used in the identification. This will be illustrated later, see Examples 4.1 and 4.2.

Finally, note that Assumption A7 means that the system operates in open loop.

2.4 SOME NOTATIONAL CONVENTIONS

In this section we will give some conventions used throughout the book.

- As has been shown in the different examples of this chapter the elements of the matrix $\phi(t)$ consist of input and output values with various time arguments. For convenience we will introduce the "noise-free" part of $\phi(t)$:

  $\tilde{\phi}(t) \triangleq \phi(t)$ where in all the elements the output y(t) (t arbitrary) is substituted by the "noise-free" output $x(t) = G(q^{-1})u(t)$, cf (2.9). As an example we get for the SISO model (2.4) of Example 2.1

  $$\tilde{\phi}^T(t) = [-x(t-1) \dots -x(t-na)\ u(t-1) \dots u(t-nb)]$$

  In view of Assumption A7 we conclude in particular that

  i) $\tilde{\phi}(t)$ depends on the input and is independent of the disturbance v(s) for all t and s

  ii) $\phi(t)-\tilde{\phi}(t)$ depends on the disturbance and is independent of the input u(s) for all t and s

- For scalar systems (ny = 1) the matrices $\phi(t)$ and $\tilde{\phi}(t)$ become (column) vectors. To emphasize that we will generally use the notation $\varphi(t)$ respectively $\tilde{\varphi}(t)$. (Cf the remark after Example 2.1).

- For polynomial degrees we will use * as superscript to denote true values and ^ to denote estimated values. For a nominal model we will thus use na, nb, and for the true system na*, nb* as in Example 2.6. If the degrees of the models are estimated from experimental data we can emphasize this by using the notation $\hat{na}$, $\hat{nb}$. This will be done in Part III of the book.

- Similar to the polynomial degree we will let θ denote the parameter vector of a nominal model while θ^* denotes the corresponding vector of the true system, see Assumptions A4, A5. Note that for a SISO system the dimension of θ^* by definition is na+nb, which may differ from (be larger than) na*+nb*. Similar differences can appear in the multivariable cases. The symbol $\hat{\theta}$ will denote an estimate of the parameter vector.

- We will occasionally compare nonnegative definite matrices. Comparison of covariance matrices for different estimates is a typical case. We will then let

$P_1 \geq P_2$

denote that P_1-P_2 is nonnegative definite.

- The noise covariance matrix Λ will be denoted λ in the scalar output case.

- The signal-to-noise ratio is denoted by S. In the scalar output case it is defined as, cf (2.9)

$$S = \sqrt{Ex^2(t)/Ew^2(t)} \tag{2.25a}$$

In the multivariable case the same definition as above applies componentwise. That is, for the i:th component of y(t) the signal-to-noise ratio is given by

$$S_i = \sqrt{Ex_i^2(t)/Ew_i^2(t)} \qquad i = 1,\dots, ny \tag{2.25b}$$

- For small remainder terms in series expansions we will use the common notations $O(x)$, $o(x)$, where

$O(x)/x$ is bounded when $x \to 0$

$o(x)/x \to 0$ when $x \to 0$

2.5 REMARKS AND BIBLIOGRAPHICAL NOTES

Assumption A1 can be extended to the case where the matrix coefficients also contain a nonzero (known) term. This means e.g. that

$$A_i(\theta) = A_{i0}+A_{i1}(\theta)$$

where $A_{i1}(0) = 0$ and the elements of $A_{i1}(\theta)$ are linear functions of the parameter vector θ. Then the general model structure (2.3) becomes

$$y(t) = \phi^T(t)\theta+d(t)+\varepsilon(t)$$

where d(t) is a known function, typically composed of lagged inputs and outputs. However, then y(t)-d(t) replaces y(t) and the analysis remains the same even if the evaluation of explicit expressions may be different. For most model structures, though, the term d(t) vanishes.

There exist several model structures or parameterizations that are not *explicitly* covered in the text.

One possibility is to let the transfer function matrix $G(q^{-1})$ be parameterized *directly*. Let $G_{ij}(q^{-1}) = B_{ij}(q^{-1})/A_{ij}(q^{-1})$. Then parameterize $G(q^{-1})$ by using the coefficients of the polynomials $B_{ij}(q^{-1})$, $A_{ij}(q^{-1})$ $i = 1, \ldots, ny$, $j = 1, \ldots, nu$ as independent parameters. The instrumental variable (IV) principle can still be applied, see e.g. Sinha and Caines (1977), Diekmann and Unbehauen (1979). However, the IV schemes based on a direct parameterization of $G(q^{-1})$ will in general give quite inaccurate estimates, Sinha and Caines (1977), unless some sophisticated ways to "generate" the instruments are used, Jakeman et al (1980), Barrett-Lenard and Blair (1981).

In the text we gave in Examples 2.4 and 2.5 two simple cases of parameterizations for MIMO models. The diagonal form has been used e.g. by Kashyap and Nasburg (1974), Sinha and Caines (1977), Gauthier and Landau (1978), El-Sherief and Sinha (1979 a, b). Two other possible parameterizations are

- The "triangular form" with the matrix polynomial $A(z,\theta)$ in Hermite form (lower left or upper right triangular), see Dickinson et al (1974), Kashyap and Nasburg (1974), Gauthier and Landau (1978), Hannan (1971).

- The "Guidorzi form" with the matrix $A(z,\theta)$ being row-proper, see e.g. Guidorzi (1975), Gauthier and Landau (1978).

These forms can also be covered by using the general model equation (2.3).

The full polynomial form of Example 2.5 has been used e.g. by Hannan (1969, 1976), Kashyap and Rao (1976), Jakeman (1979), Jakeman and Young (1979). We have analysed this form in Lemma 2.1. The equivalence i) $\Leftrightarrow$ iii) of that lemma was first proved by Hannan (1969), who also showed that, Hannan (1976), (2.19) is generically fulfilled. This form is thus what Kashyap and Rao (1976) call a pseudocanonical form. The advantage of this form is that only two structural indices, namely na and nb, need to be determined in practical applications. The second condition ii) of Lemma 2.1 is interesting from a system theoretic point of view. When establishing consistency of a certain IV method we will (in Theorem 4.4) use this condition rather than iii).

Chapter 3

BASIC AND EXTENDED IV ALGORITHMS

3.1 BASIC IV VARIANTS

In this chapter we will describe some different IV based methods for estimating the parameter vector θ in the general model, (2.3)

$$M:\ y(t) = \phi^T(t)\theta+\varepsilon(t) \tag{3.1}$$

Note that the equation error $\varepsilon(t)$ can be seen as a function of the parameter vector θ by rewriting (3.1) as

$$\varepsilon(t) = y(t)-\phi^T(t)\theta$$

For future comparison, we consider first the least squares method (LSM). Then the function

$$V(\theta) = \frac{1}{N}\sum_{t=1}^{N} \varepsilon^T(t)\varepsilon(t) = \frac{1}{N}\sum_{t=1}^{N} [y^T(t)-\theta^T\phi(t)][y(t)-\phi^T(t)\theta] \tag{3.2}$$

is minimized with respect to θ. The minimizing element is taken as the estimate $\hat{\theta}$. It is given by

$$\hat{\theta} = [\frac{1}{N}\sum_{t=1}^{N} \phi(t)\phi^T(t)]^{-1}[\frac{1}{N}\sum_{t=1}^{N} \phi(t)y(t)] \tag{3.3}$$

We next consider the consistency properties of this estimate. We say that an estimate $\hat{\theta}$ is ***consistent*** if $\hat{\theta}$ converges to the true value θ^* as the number of data, N, tends to infinity. Assume that the system S fulfils (2.10). We then get

$$\begin{aligned}\hat{\theta} &= [\frac{1}{N}\sum_{t=1}^{N} \phi(t)\phi^T(t)]^{-1}[\frac{1}{N}\sum_{t=1}^{N} \phi(t)\{\phi^T(t)\theta^*+v(t)\}] \\ &= \theta^*+[\frac{1}{N}\sum_{t=1}^{N} \phi(t)\phi^T(t)]^{-1}[\frac{1}{N}\sum_{t=1}^{N} \phi(t)v(t)]\end{aligned} \tag{3.4}$$

Under the Assumptions A2, A3, A6, A7 given in Chapter 2 the sums

$$\frac{1}{N}\sum_{t=1}^{N} \phi(t)\phi^T(t) \qquad\qquad \frac{1}{N}\sum_{t=1}^{N} \phi(t)v(t)$$

converge (with probability one) as N tends to infinity, see e.g. Söderström (1975), to the corresponding expected values. The limit of the first sum will be positive definite under weak conditions (implied by our general assumptions). We therefore find that the LS estimate (3.3) is consistent if and only if

$$E\phi(t)v(t) = 0 \tag{3.5}$$

Since $\phi(t)$ normally contains delayed output values which are correlated with the disturbances we can see that (3.5) essentially means that the disturbance $v(t)$ must be an uncorrelated sequence (i.e. white noise). This is a severe drawback of the LS method. It can also be viewed as the main reason for modifying the LS method and introducing the instrumental variable (IV) method.

The (basic) IV method for estimating θ in (3.1) is given by

$$\hat{\theta} = [\frac{1}{N}\sum_{t=1}^{N} Z(t)\phi^T(t)]^{-1}[\frac{1}{N}\sum_{t=1}^{N} Z(t)y(t)] \tag{3.6}$$

compare with (3.3). Here the matrix $Z(t)$ has dimension $n\theta|ny$. Its elements are called instruments or instrumental variables. They can be chosen in different ways. We will soon give some examples.

A calculation similar to (3.4) for the IV estimate (3.6) gives easily

$$\hat{\theta} = \theta^* + [\frac{1}{N}\sum_{t=1}^{N} Z(t)\phi^T(t)]^{-1}[\frac{1}{N}\sum_{t=1}^{N} Z(t)v(t)] \tag{3.7}$$

To get a consistent estimate we may therefore require that

$$R \triangleq EZ(t)\phi^T(t) \quad \text{is nonsingular} \tag{3.8}$$

$$0 = EZ(t)v(t) \tag{3.9}$$

In loose terms, the instruments should thus be well correlated with the lagged inputs and outputs but not correlated with the disturbance sequence. The most common attempt to fulfil these requirements is to let the instruments be delayed and possibly filtered input values. Then in light of Assumption A7 the condition (3.9) is automatically satisfied.

Before giving some examples of basic IV variants we will discuss the effect of a transformation of the instruments. Consider the IV estimate (3.6) obtained with the instruments Z(t). Assume then that Z(t) is substituted with TZ(t) where T is a $n\theta|n\theta$ nonsingular transformation matrix. With the new instruments the estimate becomes

$$\begin{aligned}\hat{\theta} &= [T\frac{1}{N}\sum_{t=1}^{N} Z(t)\phi^T(t)]^{-1}[T\frac{1}{N}\sum_{t=1}^{N} Z(t)y(t)] \\ &= [\frac{1}{N}\sum_{t=1}^{N} Z(t)\phi^T(t)]^{-1}[\frac{1}{N}\sum_{t=1}^{N} Z(t)y(t)]\end{aligned} \tag{3.10}$$

It is apparent that such a transformation does not at all affect the parameter estimate $\hat{\theta}$. This property can be used when constructing IV estimators so as to possibly simplify the computations needed. It may, of course, also be of potential use in the analysis of various IV variants.

Remark on notation. Similar to the replacement of $\phi(t)$ with $\varphi(t)$ we will use the lower case symbol z(t) for single output systems stressing that in such a case Z(t) is a (column) vector.

Next we proceed to give a few examples of basic IV variants.

Example 3.1 Some basic IV variants for SISO systems

Consider first the choice

$$z_1(t) = K(q^{-1})[-\eta(t-1)\ldots -\eta(t-na)\ u(t-1)\ldots u(t-nb)]^T \tag{3.11}$$

where $\eta(t)$ is obtained by filtering the input

$$D(q^{-1})\eta(t) = C(q^{-1})u(t) \tag{3.12a}$$

$$\begin{aligned}C(q^{-1}) &= c_0+c_1q^{-1}+\ldots+c_{nc}q^{-nc} \\ D(q^{-1}) &= 1+d_1q^{-1}+\ldots+d_{nd}q^{-nd}\end{aligned} \tag{3.12b}$$

It is assumed that $K(q^{-1})$ and $K(q^{-1})^{-1}$ are asymptotically stable filters, that D(z) has all zeros outside the unit circle and that C(z) and D(z) are coprime.

Mostly one takes nc = nb, nd = na, $K(q^{-1}) \equiv 1$. Such a case has e.g. been discussed and analysed by Finigan and Rowe (1974). With the special choice $C(z) \equiv B(z)$, $D(z) \equiv A(z)$ it has also been discussed by e.g. Wong and Polak (1967) and Young (1970 a).

We then have $z_1(t) = \tilde{\varphi}(t)$. Such a choice requires knowledge of the true system. Thus it cannot be used in a straightforward manner but some sort of adaptive or iterative techniques can be used. We will to some extent discuss this in Section 3.3 where we will allow adaptive filtering in (3.11), (3.12).

It is often possible to rewrite the IV vector (3.11). Assume that nc = nb, nd = na. Then

$$Z_1(t) = K(q^{-1}) \begin{bmatrix} -\eta(t-1) \\ \vdots \\ -\eta(t-na) \\ u(t-1) \\ \vdots \\ u(t-nb) \end{bmatrix} = \frac{K(q^{-1})}{D(q^{-1})} \begin{bmatrix} -C(q^{-1})u(t-1) \\ \vdots \\ -C(q^{-1})u(t-na) \\ D(q^{-1})u(t-1) \\ \vdots \\ D(q^{-1})u(t-nb) \end{bmatrix}$$

$$= S(-C,D)\, \frac{K(q^{-1})}{D(q^{-1})} \begin{bmatrix} u(t-1) \\ \vdots \\ u(t-na-nb) \end{bmatrix} \tag{3.13a}$$

Here $S(-C,D)$ denotes the Sylvester matrix associated with the polynomials -C and D:

$$S(-C,D) = \left[\begin{array}{ccccc} -c_0 \cdots & & -c_{nb} & & 0 \\ & \ddots & & \ddots & \\ 0 & & -c_0 \cdots & & -c_{nb} \\ \hline 1 & d_1 \cdots & & d_{na} & 0 \\ & \ddots & & & \ddots \\ 0 & & 1 & d_1 \cdots & d_{na} \end{array}\right] \tag{3.13b}$$

The rank properties of Sylvester matrices are reviewed in Lemma A3.1. In particular when C(z) and D(z) are coprime the matrix $S(-C,D)$ will be nonsingular. In view of the discussion leading to (3.10) it is thus clear that $z_1(t)$ will give exactly the same estimate as

$$z_1'(t) = \frac{K(q^{-1})}{D(q^{-1})} [u(t-1) \ldots u(t-na-nb)]^T \tag{3.13c}$$

In particular we note that the polynomial $C(q^{-1})$ has no effect at all on the estimate. To simplify the calculations we could then use (3.13c) instead of (3.11).

If we particularly take $K(q^{-1}) = D(q^{-1})$ in (3.13c) we get an IV variant described by Wouters (1972), namely

$$z_2(t) = [u(t-1) \ldots u(t-na-nb)]^T \tag{3.14}$$

A third IV variant has been proposed by Banon and Aguilar-Martin (1972), Gentil (1972) and Gentil et al. (1973). Then the instruments are given by

$$z_3(t) = [-y(t-k-1)\dots -y(t-k-na)\ u(t-1)\dots u(t-nb)]^T \tag{3.15}$$

The second consistency condition, (3.9), is then satisfied if v(t) is a moving average of order not higher than k. For nb = 0 this IV method becomes the well-known Yule-Walker procedure used for the estimation of AR parameters of a time series, see e.g. Young (1972). ■

Some other IV variants constructed to allow closed loop experiments are mentioned in the bibliographical section 3.4.

It should be stressed that although the IV variants described in Example 3.1 all satisfy the second consistency condition (3.9) it is a priori not clear under what assumptions the first consistency condition, (3.8), is fulfilled. We will have more to say on this matter in Chapter 4. We will in particular give sufficient conditions for (3.8) as well as a few examples where it is not satisfied. An idealized IV variant is discussed in the following example. Then (3.8) holds under weak conditions.

Example 3.2 *Idealized basic IV variant*

Consider the basic IV variant given by

$$Z(t) = \tilde{\phi}(t) \tag{3.16}$$

i.e. the noise-free part of $\phi(t)$, see Section 2.4. We then get in view of the discussion in Section 2.4.

$$R = E\tilde{\phi}(t)\phi^T(t) = E\tilde{\phi}(t)\tilde{\phi}^T(t) \tag{3.17}$$

Clearly R is symmetric and at least nonnegative definite. Conditions for R to be positive definite (and hence nonsingular) are basically that the model is not overparameterized (i.e. A5 holds) and that the input is persistently exciting. Exact conditions are given in Lemma 4.1.

The second consistency condition, (3.9), becomes

$$0 = E\tilde{\phi}(t)v(t)$$

which apparently is true due to Assumption A7.

It should be pointed out that this IV variant, although of theoretical interest as we will see later, cannot be used in a straightforward way. The reason is of course that $\tilde{\phi}(t)$ is not known. ∎

The IV estimate (3.6) is an off-line or batch estimate. However, it can also be implemented in an on-line form. To see this let the estimate (3.6) based on N data be denoted by $\hat{\theta}(N)$ and introduce

$$P(N) = [\sum_{t=1}^{N} Z(t)\phi^T(t)]^{-1} \tag{3.18}$$

We then have

$$P^{-1}(N) = P^{-1}(N-1)+Z(N)\phi^T(N) \tag{3.19}$$

and

$$\begin{aligned}\hat{\theta}(N) &= P(N)\sum_{t=1}^{N} Z(t)y(t) = P(N)[P^{-1}(N-1)\hat{\theta}(N-1)+Z(N)y(N)] \\ &= \hat{\theta}(N-1)+P(N)Z(N)[y(N)-\phi^T(N)\hat{\theta}(N-1)]\end{aligned} \tag{3.20}$$

Applying the matrix inversion lemma (see Lemma A3.6) to (3.19) we get finally the algorithm

$$\hat{\theta}(N) = \hat{\theta}(N-1)+K(N)[y(N)-\phi^T(N)\hat{\theta}(N-1)] \tag{3.21a}$$

$$P(N) = P(N-1)-P(N-1)Z(N)[I+\phi^T(N)P(N-1)Z(N)]^{-1}\phi^T(N)P(N-1) \tag{3.21b}$$

$$K(N) = P(N)Z(N) = P(N-1)Z(N)[I+\phi^T(N)P(N-1)Z(N)]^{-1} \tag{3.21c}$$

As long as the instruments $\{Z(t)\}$ are time-invariant and do not depend on previous estimates, the off-line estimate (3.6) and the on-line estimate (3.21) give the same result. Most of the analysis carried out in this book refers to this situation. It will then be convenient to work primarily with the off-line expression in the analysis.

One difference should be noted, though. The equation (3.21b) needs an initial value P(0) to be started. However, according to (3.18) the matrix P(0) is not defined. There are two ways to overcome this problem. One way is to start the on-line algorithm at a time $t = N_0$, chosen such that $P(N_0)$ as defined in (3.18) exists. Another way is to initialize with $P(0) = \rho I$ where $\rho >> 1$. This approach is common in recursive identification, see e.g. Ljung and Söderström (1983) for a general discussion. With the notation of (3.6) and (3.21) we then get

$$P^{-1}(N)\hat{\theta}(N) = P^{-1}(N-1)\hat{\theta}(N-1)+Z(N)y(N) = \ldots = \frac{1}{\rho}\,\hat{\theta}(0)+\sum_{t=1}^{N} Z(t)y(t)$$

$$P^{-1}(N) = \frac{1}{\rho}\,I+\sum_{t=1}^{N} Z(t)\phi^T(t)$$

and hence

$$\hat{\theta}(N)-\hat{\theta} = \frac{1}{\rho}\,P(N)\hat{\theta}(0)+P(N)[P^{-1}(N)-\frac{1}{\rho}\,I]\hat{\theta}-\hat{\theta} = \frac{1}{\rho}\,P(N)[\hat{\theta}(0)-\hat{\theta}] \qquad (3.22)$$

We see that if ρ is large then the deviation between the on-line estimate $\hat{\theta}(N)$ and the off-line estimate $\hat{\theta}$ is small.

For a further discussion of on-line identification in general including the algorithm (3.21) in particular, we refer to Ljung and Söderström (1983).

We will now end this section by discussing how the basic IV estimate should be computed for a sequence of increasing model structures.

Consider the model structure

$$M: \quad y(t) = \phi^T(t)\theta+\varepsilon(t)$$

Let $\phi^T(t)$ and θ be partitioned as

$$\phi^T(t) = [\phi_1^T(t)\ \phi_2^T(t)] \qquad \theta = [\theta_1^T\ \theta_2^T]^T$$

Consider also the smaller model structure

$$M': \quad y(t) = \phi_1^T(t)\theta_1+\varepsilon(t)$$

which can be seen as a subset of M. We illustrate the relation $M' \subset M$ by an example.

Example 3.3 Increasing model structures

Let M and M' be SISO and given by (2.4) i.e.

$$A(q^{-1})y(t) = B(q^{-1})u(t)+\varepsilon(t)$$

Further let M correspond to the polynomial degrees na,nb and M' to na',nb' with na > na', nb > nb'.

Then we have

$$\phi_1^T(t) = [-y(t-1)\dots -y(t-na')\ u(t-1)\dots u(t-nb')]$$

$$\phi_2^T(t) = [-y(t-na'-1)\dots -y(t-na)\ u(t-nb'-1)\dots u(t-nb)]$$

$$\theta_1 = [a_1\dots a_{na'}\ b_1\dots b_{nb'}]^T$$

$$\theta_2 = [a_{na'+1}\dots a_{na}\ b_{nb'+1}\dots b_{nb}]^T$$ ∎

We let the matrix of instruments Z(t) used for M be partitioned as

$$Z(t) = \begin{bmatrix} Z_1(t) \\ Z_2(t) \end{bmatrix}$$

where $Z_1(t)$ has the same dimensions as $\phi_1(t)$.

Assume now that we know

$$R = \frac{1}{N}\sum_{t=1}^{N} Z(t)\phi^T(t) = \frac{1}{N}\sum_{t=1}^{N}\begin{bmatrix} Z_1(t) \\ Z_2(t) \end{bmatrix}[\phi_1^T(t)\ \phi_2^T(t)] \triangleq \begin{bmatrix} R_{11} & R_{12} \\ R_{21} & R_{22} \end{bmatrix}$$

and the estimate

$$\hat{\theta}_1' = R_{11}^{-1}\frac{1}{N}\sum_{t=1}^{N} Z_1(t)y(t)$$

for the model structure M'. We then seek the IV estimate for the larger model structure M, i.e.

$$\hat{\theta} = R^{-1}\frac{1}{N}\sum_{t=1}^{N} Z(t)y(t)$$

Using standard equations for the inverse of a partitioned matrix we get, cf Åström (1968), Sagara et. al (1982),

$$\hat{\theta} = \begin{bmatrix} \hat{\theta}_1 \\ \hat{\theta}_2 \end{bmatrix} = \{ \begin{bmatrix} R_{11}^{-1} & 0 \\ 0 & 0 \end{bmatrix} + \begin{bmatrix} R_{11}^{-1}R_{12} \\ -I \end{bmatrix} [R_{22}-R_{21}R_{11}^{-1}R_{12}]^{-1}[R_{21}R_{11}^{-1}\ \ -I]\}\frac{1}{N}\sum_{t=1}^{N} Z(t)y(t)$$

$$= \begin{bmatrix} \hat{\theta}_1' \\ 0 \end{bmatrix} + \begin{bmatrix} -R_{11}^{-1}R_{12} \\ I \end{bmatrix} [R_{22}-R_{21}R_{11}^{-1}R_{12}]^{-1}\{\frac{1}{N}\sum_{t=1}^{N} Z_2(t)\{y(t)-\phi_1^T(t)\hat{\theta}_1'\}\}$$

These relations can be alternatively written as

$$\varepsilon(t) = y(t)-\phi_1^T(t)\hat{\theta}_1'$$

$$r = \frac{1}{N}\sum_{t=1}^{N} Z_2(t)\varepsilon(t) = \frac{1}{N}\sum_{t=1}^{N} Z_2(t)y(t)-R_{21}\hat{\theta}_1'$$

$$\hat{\theta}_2 = [R_{22}-R_{21}R_{11}^{-1}R_{12}]^{-1}r$$

$$\hat{\theta}_1 = \hat{\theta}_1'-R_{11}^{-1}R_{12}\hat{\theta}_2$$

$$R^{-1} = \begin{bmatrix} R_{11}^{-1} & 0 \\ 0 & 0 \end{bmatrix} + \begin{bmatrix} R_{11}^{-1}R_{12} \\ -I \end{bmatrix} [R_{22}-R_{21}R_{11}^{-1}R_{12}]^{-1} [R_{21}R_{11}^{-1} \quad -I]$$

Two expressions for r are given above. The first expression allows some interesting interpretations of the above relations as shown below. The second expression for r should be used for implementation since it requires a smaller amount of computations. In particular the equation errors $\varepsilon(t)$ are then not explicitly needed. If the parameter estimates are sought only for two model structures (and not a longer sequence) then the updating of R^{-1} can be dispensed with.

Note that $\varepsilon(t)$ is the equation error for the smaller model structure M'. If M' is adequate we can expect $\varepsilon(t)$ to be uncorrelated with the instruments $Z_2(t)$. In such a case r will be close to zero and the estimate for M will fulfil $\hat{\theta}_1 \approx \hat{\theta}_1'$, $\hat{\theta}_2 \approx 0$.

The scheme above is particularily useful when M contains one parameter more than M'. Then $\phi_2(t)$ and $Z_2(t)$ becomes row vectors. Further $R_{22}-R_{21}R_{11}^{-1}R_{12}$ becomes a scalar and matrix inversions can thus be completely avoided when determining the IV estimates for M.

3.2 EXTENDED IV VARIANTS

Some extensions of the basic IV estimate (3.6) will be presented in this section. The extensions include:

- prefiltering of the data
- use of an "augmented" Z(t) matrix in (3.6) so that an overdetermined system of

equations is obtained. This is then to be solved in a least squares sense.

The reasons for investigating these extensions are twofold:

- it may be easier to fulfil the consistency requirement corresponding to (3.8) for the extended IV variant. That this is indeed true for MIMO systems will be made clear in Chapter 4.

- the accuracy of the estimates may be improved if the prefilter and the instruments are chosen with care.

Consider therefore the *extended* IV estimate

$$\hat{\theta} = \arg\min_{\theta} \| [\sum_{t=1}^{N} Z(t)\cdot F(q^{-1})\phi^T(t)]\theta - [\sum_{t=1}^{N} Z(t)\cdot F(q^{-1})y(t)] \|_Q^2 \tag{3.23}$$

where $\| x \|_Q^2 = x^T Q x$, Q being a positive definite weighting matrix, and $F(q^{-1})$ is a ny|ny dimensional asymptotically stable filter. Furthermore, Z(t) is now a matrix of dimension nz|ny with nz ≥ nθ.

The optimization problem (3.23) is quadratic and can easily be solved. The estimate can be written as

$$\hat{\theta} = \{[\sum_{t=1}^{N} Z(t)\cdot F(q^{-1})\phi^T(t)]^T Q [\sum_{t=1}^{N} Z(t)\cdot F(q^{-1})\phi^T(t)]\}^{-1}$$

$$\cdot\{[\sum_{t=1}^{N} Z(t)\cdot F(q^{-1})\phi^T(t)]^T Q [\sum_{t=1}^{N} Z(t)\cdot F(q^{-1})y(t)]\} \tag{3.24}$$

From a numerical point of view it is more appropriate to find $\hat{\theta}$ for example by applying the numerically sound QR algorithm, see Stewart (1973), to the optimization problem (3.23) than to calculate the estimate as in (3.24). The expression (3.24) will, however, be useful for the theoretical analysis.

Using the system description (2.10) we get from (3.24)

$$\hat{\theta} = \theta^* + [R_N^T Q R_N]^{-1} R_N^T Q \, [\frac{1}{N}\sum_{t=1}^{N} Z(t)\cdot F(q^{-1})v(t)] \tag{3.25}$$

where

$$R_N = \frac{1}{N}\sum_{t=1}^{N} Z(t)\cdot F(q^{-1})\phi^T(t) \tag{3.26}$$

The consistency conditions (3.8), (3.9) now generalize easily to

$$\lim_{N\to\infty} R_N = R \triangleq EZ(t)\cdot F(q^{-1})\phi^T(t) \quad \text{has rank } n\theta \tag{3.27}$$

$$0 = EZ(t)\cdot F(q^{-1})v(t) \tag{3.28}$$

Note that when nz = nθ, i.e. Z(t) and φ(t) have the same dimensions, the expression (3.24) will simplify. Then there is no need to specify Q. The estimate becomes

$$\hat{\theta} = [\sum_{t=1}^{N} Z(t)\cdot F(q^{-1})\phi^T(t)]^{-1}[\sum_{t=1}^{N} Z(t)\cdot F(q^{-1})y(t)] \tag{3.29}$$

The consistency conditions (3.27)-(3.28) remain the same. Note, though, that R is now a square matrix and (3.27) simply means that R must be nonsingular.

We now proceed to give some examples.

Example 3.4 *An idealized extended IV variant*

Let the instruments be given by

$$Z(t) = [F(q^{-1})\tilde{\phi}^T(t)]^T \tag{3.30}$$

We do not specify the prefilter. Note that here Z(t) and $\tilde{\phi}(t)$ have the same dimensions. The conditions (3.27), (3.28) become, cf also Example 3.2,

$$R = E[F(q^{-1})\tilde{\phi}^T(t)]^T[F(q^{-1})\tilde{\phi}^T(t)] \quad \text{nonsingular}$$

$$0 = E[F(q^{-1})\tilde{\phi}^T(t)]^T[F(q^{-1})v(t)]$$

According to Lemma A3.7 the first condition is under weak assumptions ($F(e^{i\omega})$ nonsingular on the unit circle) equivalent to $E\tilde{\phi}(t)\tilde{\phi}^T(t)$ nonsingular. Necessary and sufficient conditions for this to be true will be given in Lemma 4.1, see also Example 3.2.

The second condition is automatically fulfilled in view of Assumption A7. ■

Example 3.5 *A class of extended IV variants for MIMO systems*

Assume that no prefiltering is used, i.e. $F(q^{-1}) \equiv I$ and that the IV matrix is block diagonal

$$Z(t) = \begin{bmatrix} z_1(t) & & \bigcirc \\ & \ddots & \\ \bigcirc & & z_{ny}(t) \end{bmatrix} \quad (3.31)$$

where the vectors $\{z_i(t)\}$ are given by

$$z_i(t) = K(q^{-1}) \begin{bmatrix} u(t-1) \\ \vdots \\ u(t-n_i) \end{bmatrix} \quad i = 1, \ldots, ny \quad (3.32)$$

and where $K(q^{-1})$ is an asymptotically stable scalar filter. For these IV variants the condition (3.28) is clearly fulfilled. Further assume that $\phi(t)$ has the diagonal structure (2.7d). The first consistency condition, (3.27) becomes

$$R_i \triangleq E z_i(t) \cdot \varphi_i^T(t) \quad \text{has rank} = \dim \varphi_i(t) \quad i = 1, \ldots, ny \quad (3.33)$$

Since R_i is a matrix of dimension $nu \cdot n_i | nai + nbi \cdot nu$ (3.33) requires that

$$nu \cdot n_i \geq nai + nbi \cdot nu = \dim \varphi_i(t) \quad i = 1, \ldots, ny \quad (3.34)$$

If the system is multivariable ($nu > 1$) and n_i is chosen as small as possible subject to (3.34), we can very well get a vector $z_i(t)$ of larger dimension than $\varphi_i(t)$. (For example, with $nai = nbi = 1$, $nu = 2$ we get $2 \cdot n_i \geq 3$. The smallest value of n_i is 2 which gives a rectangular matrix R_i of dimension $4|3$). Thus we have no longer a basic IV variant but a truly extended one. Instead of dropping some elements of the $z_i(t)$ vector in such a case it seems natural to use a $z_i(t)$ vector that is somewhat larger than absolutely necessary.

Even if we have not examined the first consistency condition for this class of IV variants (we will do so in Chapter 4), this example shows that extended IV algorithms may naturally occur when deriving consistent estimates. ∎

Similar to the treatment of basic IV methods in Section 3.1 we now consider the effects of a nonsingular linear transformation of the instruments. Clearly, such a transformation will not change the consistency properties of the estimate. The accuracy may, however, change. Indeed, consider the estimate (3.23) and let T be a nonsingular $nz|nz$ matrix. If $Z(t)$ is replaced with $T \cdot Z(t)$ then the estimate will change if $nz > n\theta$. The change can be interpreted as a modification of the weighting matrix from Q to T^TQT.

As an illustration of the above discussion we will in the following example show how this idea can be used to derive a class of IV variants. We will treat the Hammerstein model structure introduced in Example 2.3 and generalize some ideas to construct instruments commonly used in the linear SISO case (see Example 3.1).

Example 3.6 A class of IV variants for Hammerstein models

The Hammerstein model was introduced in Example 2.3. As before we use $\bar{u}(t)$ to denote the true input while

$$u(t) = [\bar{u}(t)\ \bar{u}^2(t) \ldots \bar{u}^m(t)]^T \tag{3.35}$$

will denote the auxiliary input vector, see Example 2.3.

Consider first IV vectors of the form

$$z^*(t) = K(q^{-1})[u^T(t-1) \ldots u^T(t-na-nb)]^T \tag{3.36}$$

where $K(q^{-1})$ is a scalar filter with all poles and zeros outside the unit circle.

Then introduce polynomials

$$C(z) = 1+c_1z+\ldots+c_{na}z^{na}$$

$$D_i(z) = d_{i0}+d_{i1}z+\ldots+d_{inb}z^{nb} \qquad i = 1,\ldots, m$$

such that $C(z)$ and $D_i(z)$ are coprime for all i. Introduce further the signals

$$x_i(t) = \frac{D_i(q^{-1})}{C(q^{-1})}\bar{u}^i(t) \qquad i = 1,\ldots, m$$

$$\tilde{x}(t) = \sum_{i=1}^{m} x_i(t)$$

and define

$$z^{**}(t) = [-\tilde{x}(t-1)\dots -\tilde{x}(t-na)\; u^T(t-1)\dots u^T(t-nb)\; x_2(t-1)$$
$$\dots\; x_2(t-na)\dots\; x_m(t-1)\dots\; x_m(t-na)]^T \tag{3.37}$$

We will next show that $z^{**}(t)$ can be obtained through a nonsingular linear transformation of $z^*(t)$ with $K(q^{-1}) = 1/C(q^{-1})$. This is more conveniently seen after reordering the components of $z^*(t)$. It is easy to see that

$$\begin{bmatrix} S(D_1,C) & 0 & 0 \\ 0 & \ddots & 0 \\ 0 & 0 & S(D_m,C) \end{bmatrix} \frac{1}{C(q^{-1})} \begin{bmatrix} \bar{u}(t-1) \\ \vdots \\ \bar{u}(t-na-nb) \\ \vdots \\ \bar{u}^m(t-1) \\ \vdots \\ \bar{u}^m(t-na-nb) \end{bmatrix} = \begin{bmatrix} x_1(t-1) \\ \vdots \\ x_1(t-na) \\ \bar{u}(t-1) \\ \vdots \\ \bar{u}(t-nb) \\ \vdots \\ x_m(t-1) \\ \vdots \\ x_m(t-na) \\ \bar{u}^m(t-1) \\ \vdots \\ \bar{u}^m(t-nb) \end{bmatrix} \tag{3.38}$$

The Sylvester matrices $S(D_i,C)$ of dimension $(na+nb)|(na+nb)$ are nonsingular, as $D_i(z)$ and $C(z)$ are relatively prime. The vector of the left hand side is, after a trivial reordering of the elements equal to $z^*(t)$ with $K(q^{-1}) = 1/C(q^{-1})$. Another trivial nonsingular transformation of the right hand side (mainly a reordering) gives $z^{**}(t)$.

Note that the IV method given by (3.37) can be seen as an extension of those we considered in Example 3.1 for linear SISO models.

Note also that if estimates $\hat{A}(q^{-1})$, $\hat{B}_i(q^{-1})$ of $A^*(q^{-1})$, $B_i^*(q^{-1})$ are available then it is advisable to take $D_i(q^{-1}) = \hat{B}_i(q^{-1})$, $C(q^{-1}) = \hat{A}(q^{-1})$. In this way the first $na+m\cdot nb$ components of $z^{**}(t)$ will constitute an estimate of $\tilde{\phi}(t)$, and then the matrix R may in general be better conditioned than for other choices of $C(q^{-1})$ and $D_i(q^{-1})$. Should however any pair $(\hat{A}(z), \hat{B}_i(z))$ not be (well) coprime, problems will occur. Then modifications of $C(z)$ or $D_i(z)$ become necessary. ■

3.3 BOOTSTRAP IV VARIANTS

Sometimes there are reasons to let the instruments depend on the true parameter vector θ^*. Such a case was mentioned in Example 3.2, where it was shown that the choice $Z(t) = \tilde{\phi}(t)$ fulfils the consistency condition under weak assumptions.

As θ^* is not known approximations must be used. One possibility is to use iterative algorithms. Consider again the case treated in Example 3.2.

Let $Z(t,\hat{\theta})$ denote the matrix $\tilde{\phi}(t)$ where all delayed noise-free outputs are replaced by $\hat{G}(q^{-1})u(t-j)$, with $\hat{G}(q^{-1})$ being an estimate of $G(q^{-1})$ corresponding to the parameter vector $\hat{\theta}$. Then an iterative algorithm for approximating the IV variant of Example 3.2 is the following

$$\hat{\theta}^{k+1} = [\frac{1}{N}\sum_{t=1}^{N} Z(t,\hat{\theta}^k)\phi^T(t)]^{-1}[\frac{1}{N}\sum_{t=1}^{N} Z(t,\hat{\theta}^k)y(t)] \tag{3.39}$$

To generalize the scheme to some extent consider two matrices $Z_1(t,\theta)$ and $Z_2(t,\theta)$ which both may depend on θ. An iterative estimator of IV type can then be constructed as

$$\hat{\theta}^{k+1} = [\frac{1}{N}\sum_{t=1}^{N} Z_1(t,\hat{\theta}^k)Z_2^T(t,\hat{\theta}^k)]^{-1}[\frac{1}{N}\sum_{t=1}^{N} Z_1(t,\hat{\theta}^k)y(t)] \tag{3.40}$$

When $Z_1(t,\hat{\theta}) = Z_2(t,\hat{\theta}) = \phi(t)$ the least squares estimate (3.3) is obtained. When $Z_1(t,\theta)$ does not depend on θ and $Z_2(t,\theta) = \phi(t)$ the algorithm (3.40) becomes the basic IV method (3.6).

Algorithms of the type (3.40) are sometimes called bootstraping methods. The reason is that the algorithms swap between estimating the parameter vector and the desired instruments in the following way

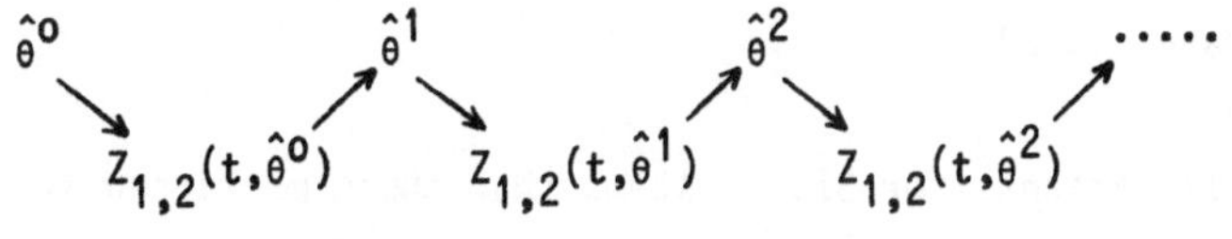

The consistency properties of the algorithm (3.40) will be examined in Section 4.4.

<u>*Example 3.7*</u> *Some bootstrap estimators*

Consider the general algorithm (3.40) for a SISO system. Let

$$\hat{\theta} = [\hat{a}_1 \ldots \hat{a}_{na}\ \hat{b}_1 \ldots \hat{b}_{nb}]^T = \text{the vector of estimated parameters} \tag{3.41}$$

$$\tilde{\varphi}(t,\hat{\theta}) = [-\frac{\hat{B}(q^{-1})}{\hat{A}(q^{-1})} u(t-1) \ldots -\frac{\hat{B}(q^{-1})}{\hat{A}(q^{-1})} u(t-na)\ u(t-1) \ldots u(t-nb)]^T \tag{3.42}$$

Consider the following two variants of a IV-based bootstrap estimator

$$BE_1: z_1(t,\hat{\theta}^k) = \tilde{\varphi}(t,\hat{\theta}^k), \qquad z_2(t,\hat{\theta}^k) = \varphi(t) \tag{3.43a}$$

$$BE_2: z_1(t,\hat{\theta}^k) = \tilde{\varphi}(t,\hat{\theta}^k), \qquad z_2(t,\hat{\theta}^k) = \tilde{\varphi}(t,\hat{\theta}^k) \tag{3.43b}$$

The motivations for these two estimators are as follows: Both BE_1 and BE_2 can be seen as iterative ways of implementing the idealized IV estimator with instruments $Z(t) = \tilde{\phi}(t)$ which we discussed in Example 3.2. Clearly, $z_1(t,\hat{\theta}^k) = \tilde{\varphi}(t,\hat{\theta}^k)$ will approximate $\tilde{\varphi}(t)$. The approximation will be accurate if the estimate $\hat{\theta}^k$ is close to the "true value" θ^*. The second algorithm, BE_2, can be seen as a symmetric version of BE_1, since the matrix inverse appearing in (3.40) is always symmetric for BE_2. It can also be seen as a way to approximately implement the idealized IV estimator of Example 3.2 if we write the system in the form

$$y(t) = \tilde{\varphi}^T(t)\theta^* + \text{noise}$$ ■

3.4 REMARKS AND BIBLIOGRAPHICAL NOTES

The LS method can also be defined in slightly different ways than we did in (3.1) - (3.3). Let $h(Q)$ be an arbitrary monotonically increasing function with positive definite matrices Q as domain. Then $\hat{\theta}$ as given by (3.3) minimizes $h(\frac{1}{N}\sum_1^N \varepsilon(t)\varepsilon^T(t))$, see Söderström and Stoica (1980). In (3.2) we used the common case $h(Q) = \text{tr}\ Q$.

The IV variant (3.12) with $K(q^{-1}) = 1$, $C(z) = B(z)$, $D(z) = A(z)$ has also been discussed independently by Mayne (1967).

There certainly exist other IV variants for SISO systems than those mentioned in Example 3.1, although (3.11), (3.12) cover the most common cases. The instrumental vector (3.15) can e.g. be generalized to

$$z(t) = [-y(t-k-1) \ldots -y(t-k-na)\ u(t-k-1) \ldots u(t-k-nb)]^T$$

The advantage of this IV variant, see Stoica and Söderström (1979) for a more detailed discussion, is that the system can be allowed to operate either in open or closed loop. We still have to assume that the disturbance v(t) is a moving average of an order not exceeding k.

Another IV variant which is directly designed for systems operating under feedback has been proposed by Bauer and Unbehauen (1978). Assume that the feedback is given by

$$u(t) = r(t)-M(q^{-1})y(t)$$

where r(t) is a persistently exciting setpoint and $M(q^{-1})$ a filter describing the feedback dynamics. Then we can take the instruments as

$$z(t) = K(q^{-1})[-\eta(t-1)\ldots -\eta(t-na)\ r(t-1)\ldots\ r(t-nb)]^T$$

$$D(q^{-1})\eta(t) = C(q^{-1})r(t)$$

cf (3.11) - (3.12). The properties of this IV variant are discussed by Söderström and Stoica (1981 a).

More aspects on recursive identification methods in general, including the recursive IV algorithm (3.21) are discussed in Ljung and Söderström (1983).

The idea of using prefilters (one of the two ideas leading to extended IV) has been proposed by Young (1970 b) in the context of IV methods. The way of using $F(q^{-1})$, Q and a augmented z(t) as (additional) design variables is based on our former work, Stoica and Söderström (1981 d, 1982 d, 1983). Similarly, Example 3.6 is based on Stoica and Söderström (1982 a).

Bootstrap estimators, such as those discussed in Example 3.7 have been proposed by Rowe (1970) and Pandya (1972, 1974).

IV variants with a augmented IV vector ($nz > n\theta$) have been obtained using correlation techniques e.g. by Peterka and Halousková (1970), Isermann and Baur (1974), Stoica and Stoica (1976), Haber (1979).

Extended IV estimates can, similarly to the basic IV estimate, be implemented as an on-line algorithm. The resulting equations are more complex than (3.21), see Friedlander (1982 b) for details.

We end this chapter by giving a table which surveys several references dealing with IV estimation. Needless to say, even if it is extensive we do not claim that the table includes every single reference in the field. Note that in the table by

REFERENCE	SYSTEM		MODEL STRUCTURE				IV VARIANT				ANALYSIS		
	scalar	multivariable	SISO difference eqn.	MIMO full pol. form	MIMO diagonal form	MIMO direct parameteriz.	basic	extended	bootstrap	optimal	consistency	accuracy	model structure determination
Ahmed (1982)		x		x 1)			x						x
Andeen and Shipley (1963)	x		x				x						
Banon and Aguilar-Martin (1972)	x		x				x				x		
Barrett-Lenard and Blair (1981)		x				x			x				
Bauer and Unbehauen (1978)	x		x				x		x				
van den Boom (1982)	x		x				x	x	x				x
Cadzow (1980)	x		x 2)					x					
Cadzow (1982)	x		x 2)				x	x					
Caines (1976 b)	x		x				x					x	
Chan (1973)		x		x			x						
Chan and Langford (1982)	x		x 2)				x						
Dhrymes et al (1970)	x		x				x			x	x		
Diekmann and Unbehauen (1979)		x				x			x				
Engle (1980)	x		x							x		x	
Eykhoff (1980)	x		x				x						
Finigan (1976)	x		x				x		x		x		
Finigan and Rowe (1973)	x		x				x						
Finigan and Rowe (1974)	x		x				x				x		
Friedlander (1982 a)	x		x 2)				x		x				
Friedlander (1982 b)	x		x 2)				x	x					
Gauthier and Landau (1978)		x		x 1)	x				x				
Gentil et al (1973)	x		x				x						
Gersch (1970)	x		x 2)				x				x	x	
Hannan (1975)		x		x 2)			x				x		
Hausman (1975)		x		x 1)						x		x	
Isermann and Baur (1974)	x		x					x					
Isermann et al (1974)	x		x					x	x				
Jakeman (1979)		x		x					x	x			
Jakeman et al (1980)		x				x			x	x			
Jakeman and Young (1979)		x		x				x		x			
Jakeman and Young (1981)	x		x					x		x			

Table 3.1. Survey of the literature on IV estimation of linear dynamic systems.

1) The model structure is a structural polynomial form
2) The model structure is an AR model (a pure time series is treated).

REFERENCE	SYSTEM		MODEL STRUCTURE				IV VARIANT				ANALYSIS		
	scalar	multivariable	SISO difference eqn.	MIMO full pol. form	MIMO diagonal form	MIMO direct parameteriz.	basic	extended	bootstrap	optimal	consistency	accuracy	model structure determination
Jorion and Hanus (1980)	x		x				x		x				
Joseph et al (1961)	x		x				x						
de Keyser (1979)	x		x				x		x				
Kim and Cain (1982)	x		x				x				x		
Ljung and Trulsson (1981)	x		x				x		x				
Ljung and Söderström (1983)	x	x	x	x 1)			x	x	x	x	x	x	
Lyttkens (1974)		x		x 1)			x		x	x	x	x	
Mayne (1967)	x		x 2)						x				
Mehra (1971)	x		x				x						
Pandya (1972)	x		x						x				
Pandya (1974)	x		x						x				
Pandya and Pagurek (1973)	x		x						x				
Peterka and Halousková (1970)	x		x					x					
Peterka and Šmuk (1969)	x		x					x					
de la Puente and Albertos (1979)	x		x				x						
Rowe (1970)		x		x 1)					x		x		
Sagara and Wada (1977)	x		x				x				x		
Sagara et al (1982)	x		x				x						x
Sinha and Caines (1977)		x				x	x						
Sinha and Sen (1975)	x		x						x				
Söderström (1974)	x		x				x				x		
Söderström et al (1974 a)	x		x				x		x				
Söderström et al (1978)	x		x				x		x				
Söderström and Stoica (1978)	x		x				x			x	x	x	
Söderström and Stoica (1979 a)	x		x							x		x	
Söderström and Stoica (1979 b)	x		x				x				x	x	
Söderström and Stoica (1981 a)	x		x				x				x	x	x
Stoica (1982 a)		x		x 2)				x			x		x
Stoica and Stoica (1976)	x		x					x					
Stoica and Söderström (1979)	x		x				x				x		

Table 3.1. Continued.

REFERENCE	SYSTEM		MODEL STRUCTURE				IV VARIANT				ANALYSIS		
	scalar	multivariable	SISO difference eqn.	MIMO full pol. form	MIMO diagonal form	MIMO direct parameteriz.	basic	extended	bootstrap	optimal	consistency	accuracy	model structure determination
Stoica and Söderström (1981 a)	x		x						x		x	x	
Stoica and Söderström (1981 b)	x		x					x		x	x	x	
Stoica and Söderström (1981 d)	x	x	x	x	x		x	x		x	x	x	x
Stoica and Söderström (1982 a)		x			x		x	x			x		
Stoica and Söderström (1982 b)	x		x							x		x	
Stoica and Söderström (1982 c)		x		x 1)	x		x	x			x	x	
Stoica and Söderström (1982 d)		x		x						x		x	
Stoica and Söderström (1983)	x		x					x		x	x	x	
Tzafestas (1970)		x		x					x				
Ward (1977)	x		x				x			x	x	x	
Wellstead (1978)	x		x				x						x
Wellstead and Rojas (1982)	x		x				x						x
Whitehead and Young (1979)		x		x 1)				x					
Whitehead et al (1979)	x		x					x					
Wong and Polak (1967)	x		x				x		x	x	x	x	
Wouters (1972)	x		x				x						
Young (1965 a)	x		x				x						
Young (1965 b)	x		x				x						
Young (1968)	x		x						x				
Young (1970 a)	x		x				x		x				
Young (1970 b)	x		x 2)				x		x	x			
Young (1972)	x		x				x						
Young (1974)	x		x				x	x					
Young (1976)	x		x						x	x			
Young (1981)	x		x				x	x		x			
Young and Hastings-James (1970)	x		x				x						
Young and Jakeman (1979 a)	x		x					x		x			
Young and Jakeman (1979 b)	x	x	x	x				x	x	x			
Young and Jakeman (1980)	x	x	x	x		x		x		x			
Young et al (1980)	x		x						x				x
Young et al (1971)	x		x						x				
Young and Whitehead (1977)		x		x					x				

Table 3.1. Continued.

"analysis" we refer to *theoretical* examinations. Simulations are not reported. No doubt, simulation studies are good complements to the theory when investigating the properties of a method. Extensive Monte Carlo simulations for investigating the properties of IV methods can be found e.g. in Jakeman (1979), Jakeman et al. (1980), Jakeman and Young (1979, 1981), Young and Jakeman (1979 a). We may also remark that some information on applications of IVMs to real data can be found in Table 9.2 of Chapter 9.

PART II

ANALYSIS

Chapter 4

CONSISTENCY

4.1 GENERAL CONSIDERATIONS

In this chapter we will examine the consistency properties of the various IV estimators introduced in Chapter 3. In this first section we will make some general remarks. In particular we shall discuss what necessary conditions we have to impose on the model structure $\mathcal{M}$ and the experimental condition X. We will also in this section present a general result concerning the "generic" consistency of IVMs.

We have briefly made some consistency considerations already in Chapter 3. We then said that $\hat{\theta}$ is a consistent estimate of θ^* if $\hat{\theta}$ converges to θ^* as N tends to infinity. Now, $\hat{\theta}$ is a stochastic variable and then several convergence concepts are possible, Chung (1968). When dealing with consistency we will use "convergence with probability one" (w.p.1). This means that the probability for the event ($\hat{\theta} \to \theta^*$ as $N \to \infty$) is 1. A weaker concept (implied by, but not equivalent to convergence w.p.1) is "convergence in probability". Then it is required that $P(|\hat{\theta}-\theta^*| > \varepsilon) \to 0$, $N \to \infty$ for every $\varepsilon > 0$. A still weaker but useful concept is "convergence in distribution". We will prove in Chapter 5 that under suitable assumptions $\sqrt{N}(\hat{\theta}-\theta^*)$ converges in distribution to a gaussian variable, i.e. the distribution density of $\sqrt{N}\,(\hat{\theta}-\theta^*)$ converges to the gaussian distribution function. We then will use the phrasing $\sqrt{N}\,(\hat{\theta}-\theta^*)$ is "asymptotically gaussian distributed".

Consider first the basic IV estimate (3.6). It clearly fulfils the equation (3.7). Assuming that i) the limits

$$R = \lim_{N\to\infty} \frac{1}{N} \sum_{t=1}^{N} Z(t)\phi^T(t)$$
$$r = \lim_{N\to\infty} \frac{1}{N} \sum_{t=1}^{N} Z(t)v(t) \qquad (4.1)$$

exist and ii) R is nonsingular, it follows readily that

$$\lim_{N\to\infty} \hat{\theta} = \theta^* + R^{-1} r \qquad (4.2)$$

We can very well use limit with probability one in (4.1)-(4.2), see Söderström (1975) and Appendix 4. If the weaker concept "limit in probability" is used, (4.2) follows from Slutzky's theorem, see Lemma A4.3. We thus get consistency (i.e $\lim_{N\to\infty} \hat{\theta} = \theta^*$) if R is nonsingular and r = 0. Under the general assumptions made in Chapter 2 it follows e.g. from Söderström (1975) that the limits in (4.1) are equal to the corresponding expected values.

The analysis can easily be performed also for the extended IV variants. We then get instead of (4.2), cf (3.25)

$$\lim_{N\to\infty} \hat{\theta} = \theta^* + (R^T Q R)^{-1} R^T Q r \tag{4.3}$$

$$R = EZ(t)\cdot F(q^{-1})\phi^T(t) \tag{4.4a}$$

$$r = EZ(t)\cdot F(q^{-1})v(t) \tag{4.4b}$$

With very few exceptions (notably z_3 in Example 3.1) the instruments are constructed using lagged and possibly filtered inputs. Since outputs are not used and the input is independent of the disturbances (Assumption A7) we get trivially r = 0. We formalize this as an additional assumption, which we will generally use in the analysis.

A8	The instruments Z(t) and the disturbance v(s) are independent for all t and s

Table 4.1 Assumption on the IV method

As a consequence of Assumption A8 we can substitute $\phi(t)$ in (4.4a) with its "noise-free" part $\tilde{\phi}(t)$, cf Section 2.4. We thus have

$$R = EZ(t)\cdot F(q^{-1})\tilde{\phi}^T(t) \tag{4.5}$$

The main consistency condition is then, cf (4.3) (also, see (3.27))

$$\text{rank } R = n\theta \tag{4.6}$$

In this section we will discuss three conditions that are necessary for (4.6) to be true:

i) $nz \geq n\theta$. This is a condition on the identification method.

ii) The input u(t) must be persistently exciting (p.e.) of an appropriate order. This is a requirement on the experimental condition X. See Assumption A6.

iii) Assumption A5 (θ^* exists and is unique) is satisfied. For a given S this is a condition on the parameterization, i.e. on the model structure M.

The first condition is trivial, since the matrix R is of dimension $nz|n\theta$ and can never have a rank that exceeds nz.

To realize the second condition think of an extreme case. Let the input u(t) be a constant. Then, in the stationary phase, x(t) will also be a constant as well as Z(t) and $\tilde{\phi}(t)$. The matrix R then becomes

$$R = Z \cdot F(1) \cdot \tilde{\phi}^T \tag{4.7}$$

Since F(1) is a $ny|ny$ dimensional matrix we get rank $R \leq ny$. However, in all practical cases we have $ny < n\theta$ and thus (4.6) cannot be fulfilled. This example shows the need to require persistent excitation. However, it is not trivial to know of what order the input must be p.e. In fact, this will depend on the model structure selected. We will return to this point in Examples 4.1 and 4.2 where we will discuss it in some detail for SISO and MISO systems.

We will now show that it is in fact necessary for consistency that the covariance matrix

$$\tilde{R} = E\tilde{\phi}(t)\tilde{\phi}^T(t) \tag{4.8}$$

is nonsingular. This is not added as a fourth condition since we will show in Lemma 4.1 below that it is equivalent to ii) and iii). Note that for the specific IV variant we discussed in Example 3.2 we had to assume that $\tilde{R}$ is nonsingular. The proof that the matrix $\tilde{R}$ must be nonsingular is by contradiction. Assume that $\tilde{R}$ is singular. Then there is a nonzero vector $\bar{r}$ such that

$$0 = \bar{r}^T\tilde{R}\bar{r} = E[\tilde{\phi}^T(t)\bar{r}]^T[\tilde{\phi}^T(t)\bar{r}]$$

or equivalently

$$\tilde{\phi}^T(t)\bar{r} = 0 \quad \text{w.p.1}$$

We then get

$$R\bar{r} = EZ(t)\cdot F(q^{-1})\tilde{\phi}^T(t)\bar{r} = 0$$

which shows that R must have rank less than $n\theta$. Thus we can conclude that $\tilde{R}$ nonsingular is a necessary condition for (4.6). We continue to show that this condition is equivalent to ii) and iii). This is done in the following lemma.

Lemma 4.1 Consider the system (2.9) and the model structure (2.3). Let Assumptions

A2 (a linear finite order system) and

A4 (existence of a true parameter vector θ^*)

be satisfied. Then the following two assertions are equivalent

i) $\tilde{R}$, (4.8), is positive definite

ii) u(t) is persistently exciting and Assumption A5 (θ^* is unique) holds.

Proof. The lemma follows from the following series of equivalences. They all hold with probability one. The explanations for the equivalences are given within brackets.

$$r^T\tilde{R}r = 0 \Leftrightarrow E\|\tilde{\phi}^T(t)r\|^2 = 0 \Leftrightarrow$$

$$\tilde{\phi}^T(t)r = 0 \Leftrightarrow \{\text{note that } x(t) = \tilde{\phi}^T(t)\theta^*\}$$

$$x(t) = \tilde{\phi}^T(t)[\theta^*-r] \Leftrightarrow \{\text{cf (2.9), A4 and (2.10)}\}$$

$$A(q^{-1},\theta^*-r)x(t) = B(q^{-1},\theta^*-r)u(t) \Leftrightarrow \{\text{use (2.9), A4}\}$$

$$[A(q^{-1},\theta^*-r)^{-1}B(q^{-1},\theta^*-r)-A(q^{-1},\theta^*)^{-1}B(q^{-1},\theta^*)]u(t) = 0 \Leftrightarrow \{u(t) \text{ p.e.}\}$$

$$[A(q^{-1},\theta^*-r)^{-1}B(q^{-1},\theta^*-r)-A(q^{-1},\theta^*)^{-1}B(q^{-1},\theta^*)] \equiv 0 \Leftrightarrow \{\text{A5}\}$$

$$r = 0$$

∎

The result of Lemma 4.1 is somewhat vague, since it is not specified of what order the input must be persistently exciting. To get more insight we will investigate the case of SISO as well as MISO systems in some detail.

Example 4.1 Analysis of nonsingularity of $\tilde{R}$ for a SISO model

Consider the SISO model (2.4) introduced in Example 2.1. To analyse the nonsingularity of $\tilde{R}$ consider the equation

$$\tilde{\phi}^T(t)r = 0 \quad \text{w.p.1} \tag{4.9}$$

and write r as

$$r = [\tilde{a}_1 \ldots \tilde{a}_{na} \; \tilde{b}_1 \ldots \tilde{b}_{nb}]^T$$

Then (2.4d), (4.9) give

$$[-\tilde{A}(q^{-1}) \frac{B^*(q^{-1})}{A^*(q^{-1})} + \tilde{B}(q^{-1})]u(t) = 0 \tag{4.10}$$

where

$$\tilde{A}(z) = \tilde{a}_1 z + \ldots + \tilde{a}_{na} z^{na}$$

$$\tilde{B}(z) = \tilde{b}_1 z + \ldots + \tilde{b}_{nb} z^{nb}$$

Assume first that u(t) is p.e. of order $\bar{n} \triangleq \max(na+nb^*, na^*+nb)$. According to Result A1.5 the same is true for $\frac{1}{A^*(q^{-1})}$ u(t), and then (4.10) implies (see, for example, Result A1.5)

$$-\tilde{A}(z)B^*(z)+A^*(z)\tilde{B}(z) = 0$$

The analysis of this equation parallels the discussion in Example 2.6. Since A* and B* are coprime we get

$$\tilde{A}(z) = \tilde{B}(z) = 0 \quad \text{if } n^* \triangleq \min(na-na^*, nb-nb^*) = 0$$

$$\tilde{A}(z) = A^*(z)L(z), \; \tilde{B}(z) = B^*(z)L(z), \; L(z) = \ell_1 z + \ldots + \ell_{n^*} z^{n^*} \quad \text{if } n^* > 0$$

If u(t) is not p.e. of order $\bar{n}$, then there is a polynomial M(z) of degree $\bar{n}$ such that(4.10) implies

$$-\tilde{A}(z)B^*(z)+A^*(z)\tilde{B}(z) \equiv M(z)$$

This identity is to be solved with respect to the coefficients of $\tilde{A}(z)$, $\tilde{B}(z)$. It has a unique (and nontrivial) solution if $n^* = 0$ and an infinite number of solutions when $n^* > 0$.

To summarize, we see that (4.9) implies $r = 0$ (i.e. $\tilde{R}$ is positive definite) if and only if

i) u(t) is p.e. of order max(na+nb*, na*+nb)

ii) $n^* = \min(na-na^*, nb-nb^*) = 0$

The first condition means that the input must be rich enough to excite all modes of the system. This condition is not so precise in the general case given in Lemma 4.1. The second condition states that the model is not overparameterized, so that no harmful pole-zero cancellation can take place. This latter condition was shown in Example 2.6 to be equivalent to A5. ∎

Next we study the nonsingularity of $\tilde{R}$ for a MISO model. The result will be slightly different from that derived in Example 4.1.

<u>*Example 4.2*</u> *Analysis of nonsingularity of* $\tilde{R}$ *for a MISO model*

Consider the MISO model (2.5) introduced in Example 2.2. We obtain similarly to (3.13a)

$$\tilde{\varphi}(t) = \bar{S}(-B^*, A^*)\psi(t)$$

where $\bar{S}(-B^*, A^*)$ is a generalized Sylvester matrix of dimension $(na+nb\cdot nu)|nu\cdot(na+nb)$, see Definition A3.2, and

$$\psi(t) = \frac{1}{A^*(q^{-1})}\begin{bmatrix} u(t-1) \\ \vdots \\ u(t-na-nb) \end{bmatrix}$$

Note that $\bar{S}(-B^*, A^*)$ is a rectangular matrix for $nu > 1$.

We then have

$$\tilde{R} = E\tilde{\varphi}(t)\tilde{\varphi}^T(t) = \bar{S}(-B^*, A^*)[E\psi(t)\psi^T(t)]\bar{S}^T(-B^*, A^*)$$

It follows from Lemma A3.2 that

$$n^* = \min(na-na^*, nb-nb^*) = 0$$

is a necessary condition for $\tilde{R}$ to be nonsingular.

To derive a sufficient condition it is most convenient to require in addition that $E\psi(t)\psi^T(t)$ is positive definite. This means according to Result A1.5 that u(t) is persistently exciting of order na+nb (= max(na*+nb, na+nb*) for n* = 0).

The above requirement on u(t) is, however, not strictly necessary. What is precisely required is the following condition:

Let $\rho \neq 0$ be an arbitrary vector in the nullspace of $E\psi(t)\psi^T(t)$. Then the linear system $\bar{S}^T(-B^*, A^*)\xi = \rho$ (which is over-determined!) must not have any solutions.

Clearly, the above condition is not very practical. For example, it depends on the unknown system! Instead it is convenient to use the (somewhat stronger) assumption of persistent excitation. Note that when u(t) is p.e. of order na+nb then no vector $\rho \neq 0$ will exist. ■

We have so far studied some necessary conditions for R to have full rank. We will see in the following discussions that in the "normal" case R will have full rank (provided the necessary conditions are satisfied) but also that some rare cases exist where R has lower rank. In order to formalize such a result we have to define the meaning of the concept "generically true". It is often used also in multivariable systems theory, see Wonham (1974). It is connected to the Lebesgue measure for which we refer to Cramér (1946) and Pearson (1974).

Definition 4.1 Let Ω be an open set. We then say that a statement s is *generically true with respect to* $\rho \in \Omega$ if the set

$$M = \{\rho | \rho \in \Omega, s \text{ is not true}\}$$

has Lebesgue measure zero in Ω. ■

The Lebesgue measure is discussed in Appendix 2. In more loose terms, we can say that we require M to have smaller dimension than Ω. If s is generically true with respect to $\rho \in \Omega$ and we choose a $\rho \in \Omega$ randomly then the probability that $\rho \in M$ is zero, i.e. the probability that s is true is one. In particular if s is true for all $\rho \in \Omega$ it is trivially generically true with respect to $\rho \in \Omega$.

A main result of this chapter is that rank $R = n\theta$ is generically true under weak conditions. We will now formalize this as a general theorem. Later we will apply it to some specific IV variants.

Theorem 4.1 Consider the matrix R, (4.5). Let it depend on a finite dimensional vector ρ which belongs to the open connected set Ω. Assume that

i) The elements of R are analytic functions of every element of $\rho \in \Omega$.

ii) There is a vector $\rho^* \in \Omega$ such that $R(\rho^*)$ has rank equal to $n\theta$.

Then rank $R(\rho) = n\theta$ is generically true with respect to $\rho \in \Omega$.

Proof. The result follows immediately from Lemma A2.3 and its corollary. ■

We next discuss the first assumption of the theorem, while assumption ii) will be verified for some typical IV variants in the following sections.

Example 4.3 *A possible parameter vector* ρ

Consider the SISO model (2.4a) and the IV variant (3.11)-(3.12a). Assume that $K(q^{-1})$ is a finite order (rational) filter specified by the parameters $k_1 \dots k_{nk}$. Further, let the input spectral density be specified by the parameters $\nu_1 \dots \nu_{n\nu}$. Let for example, the input u(t) have a rational spectral density. Then $\{\nu_i\}$ could be the coefficients of the ARMA representation of u(t).

We now may take

$$\rho = [a_1^* \dots a_{na*}^* b_1^* \dots b_{nb*}^* k_1 \dots k_{nk}\ c_0 \dots c_{nc}\ d_1 \dots d_{nd}\ \nu_1 \dots \nu_{n\nu}]^T$$

The set Ω is the subset of $R^{na*+nb*+nk+nc+nd+n\nu}$ given by the constraints

A*(z), K(z), $K^{-1}(z)$ and D(z) have all zeros strictly outside the unit circle
A*(z) and B*(z) are coprime
C(z) and D(z) are coprime

Additional and similar constraints may be added on $\nu_1 \dots \nu_{n\nu}$, depending on how these parameters are introduced.

Under these conditions the elements of R will be analytic functions of every element of ρ for $\rho \in \Omega$. The first condition of Theorem 4.1 is thus satisfied.

It is clear that this example can be generalized in a straightforward way to the multivariable case. The parameter vector ρ can then consist of the true values θ^*, filter parameters used to create the instruments Z(t) from the measured data, and parameters describing the input spectral density. ∎

The implications of Theorem 4.1 are threefold. First, from a practical viewpoint there is no need to worry about rank-deficient R matrices since they appear so seldom (provided some mild conditions are fulfilled). Second, from a theoretical point of view the counterexamples to general consistency that have a rank-deficient R matrix should be interesting. We will give some examples of this kind later, see Example 4.4. Third, the matrix R may be ill-conditioned (nearly rank-deficient). This will happen "close" to the set where R is rank deficient. Therefore R can be ill-conditioned in a set of non-zero Lebesgue measure. As we will see in Chapter 5 the asymptotic distribution of $\hat{\theta}$ will then have a large covariance matrix. It is therefore likely that the parameter estimate $\hat{\theta}$ then will become quite inaccurate.

4.2 ANALYSIS OF IV VARIANTS FOR SCALAR SYSTEMS

In this section we will give sufficient conditions for consistency of some IV variants for SISO systems. These IV variants were introduced in Example 3.1. According to the discussion in Section 4.1 it is sufficient to give conditions that guarantee that (4.6) is fulfilled.

In the analysis we will make repeated use of Sylvester matrices, which are analysed in Appendix A3.

We will also make use of the following covariance matrix

$$P(D,A,u,m_1,m_2) = E \begin{bmatrix} \frac{1}{D(q^{-1})} u(t-1) \\ \vdots \\ \frac{1}{D(q^{-1})} u(t-m_1) \end{bmatrix} \left[\frac{1}{A(q^{-1})} u(t-1) \dots \frac{1}{A(q^{-1})} u(t-m_2) \right] \quad (4.11)$$

where it is assumed that A(z) is a polynomial of degree na and with all zeros outside the unit circle. It is also assumed that 1/D(z) is a rational function with all poles and zeros outside the unit circle. Conditions for the matrix (4.11) to be nonsingular are given in Lemma A3.8.

We can now proceed to analyse the consistency properties of the basic IV variant given by (3.11), i.e. when

$$z(t) = K(q^{-1})[-\eta(t-1)\dots -\eta(t-na)\ u(t-1)\dots u(t-nb)]^T \tag{4.12}$$

where the auxiliary signal $\eta(t)$ is obtained by filtering the input:

$$D(q^{-1})\eta(t) = C(q^{-1})u(t) \tag{4.13a}$$

$$\begin{aligned} C(q^{-1}) &= c_0+c_1q^{-1}+\dots+c_{nc}q^{-nc} \\ D(q^{-1}) &= 1+d_1q^{-1}+\dots+d_{nd}q^{-nd} \end{aligned} \tag{4.13b}$$

As mentioned in Section 3.1 the specific IV variant given by (3.14) is a special case of (4.12), (4.13) (take $K(q^{-1}) = 1$, $C(q^{-1}) = q^{-nb}$, $D(q^{-1}) = 1$).

We now proceed to give sufficient conditions for consistency of the IV method with instruments given by (4.12)-(4.13).

<u>*Theorem 4.2*</u> Consider the IV method given by (3.6), (4.12), (4.13). Suppose that Assumptions

A2, A4, A5 (existence of an asymptotically stable system corresponding to a unique θ^*)

A3, A6, A7 ($u(\cdot)$, $v(\cdot)$ stationary and independent)

all apply. Assume further that

i) $K(q^{-1})$ and $K(q^{-1})^{-1}$ are asymptotically stable

ii) nc = nb, nd = na

iii) C(z), D(z) are coprime and have all zeros outside the unit circle.

Then the consistency condition (3.8) is fulfilled if either of the following two conditions is true:

I) u(t) is white noise

II) $K(q^{-1})A^*(q^{-1})/D(q^{-1})$ is a strictly positive real filter.

Proof. Some straightforward calculations give

$$R = Ez(t)\varphi^T(t) = Ez(t)\tilde{\varphi}^T(t)$$

$$= S(-C,D)P(D/K,A^*,u,na+nb,na+nb)S^T(-B^*,A^*) \tag{4.14}$$

where all matrices are square of dimension (na+nb)|(na+nb). Here, we have used the system description (2.10), (2.13a). Also note that Assumption A5 implies

$$n^* = \min(na-na^*, nb-nb^*) = 0$$

and $A^*(z)$, $B^*(z)$ coprime, see (2.13), (2.15). It then follows from Lemma A3.1 that $S(-C,D)$ and $S^T(-B^*,A^*)$ are nonsingular. Finally it follows from Lemma A3.8 that the middle matrix P is nonsingular if i), and I) or II) are satisfied. This completes the proof. ∎

Remark 1. Note that if $K(q^{-1}) = 1$, $na^* = 1$, $nd = 1$ then the positive realness condition is always fulfilled. This is seen as follows. Noting that $|a^*| < 1$, $|d| < 1$ we get

$$\mathrm{Re}\,\frac{K(e^{i\omega})A^*(e^{i\omega})}{D(e^{i\omega})} = \mathrm{Re}\,\frac{A^*(e^{i\omega})D(e^{-i\omega})}{D(e^{i\omega})D(e^{-i\omega})} = \frac{1+a^*d+(a^*+d)\cos\omega}{1+d^2+2d\cos\omega} \geq \frac{(1-|a^*|)(1-|d|)}{(1+|d|)^2} > 0$$

∎

Remark 2. Note that it follows from Lemma A3.8 that the result of the theorem remains true also if u(t) is an AR process of an order not exceeding na-na*+nb.

∎

Remark 3. Part (II) suggests a way of choosing the filters. The choice $D(q^{-1}) = A^*(q^{-1})$, $K(q^{-1}) = 1$ clearly gives strictly positive realness. Note that then the idealized IV variant introduced in Example 3.2 is obtained. It is therefore appealing to take $K(q^{-1}) = 1$ and to make an adaptive filtering, $D(q^{-1}) = \hat{A}(q^{-1})$, with the $\hat{A}(q^{-1})$ polynomial obtained from actual parameter estimates assuming a recursive or iterative processing of the data. This idea has been used, perhaps for other reasons, in several papers, e.g. Wong and Polak (1967), Young (1970 a) and Pandya (1972). It will be further discussed in Section 4.4. Also note that a similar condition of strictly positive realness and the potential to use it for an adaptive choice of $D(q^{-1})$ appears in some model reference schemes, cf Landau (1974, 1979). Finally note that for the IV variant given by (3.14) condition II becomes: $A^*(q^{-1})$ strictly positive real. ∎

For the other IV variants mentioned in Section 3.1 only quite restricted sufficient consistency conditions are known. We refer to Söderström and Stoica (1981 a) for details. Note, though, that in the light of Theorem 4.1 these even restricted results can be used to prove generic consistency.

It follows from Theorem 4.1 that the consistency condition (4.6) is *generically true* if in Theorem 4.2 Assumptions A2-A7, i) and iii) are satisfied, and ii) is substituted with the more general assumption $nc \geq nb$, $nd \geq na$. These are quite general assumptions. They are almost identical to the necessary conditions we discussed in Section 4.1. If the filters $K(q^{-1})$, $C(q^{-1})/D(q^{-1})$ and $B^*(q^{-1})/A^*(q^{-1})$ are picked at random, subject to natural stability constraints then for a given p.e. input signal the probability for rank $R = n\theta$ is one. (cf Theorem 4.1 and sufficiency condition II of Theorem 4.2).

Similarly, for given filters $K(q^{-1})$, $C(q^{-1})/D(q^{-1})$ and $B^*(q^{-1})/A^*(q^{-1})$ we can randomly choose the parameters of an ARMA input process and still have rank $R = n\theta$ with probability one. (cf Theorem 4.1 and sufficient condition I of Theorem 4.2).

As mentioned previously there exist though cases where the matrix R does not have full rank. We now illustrate this fact with the following example.

Example 4.4 *Counterexamples to general consistency for basic IV variants*

Consider the basic IV variant with instruments defined according to (4.12), (4.13). Assume that the general assumptions apply, and let $K(q^{-1}) = 1$, $na = na^* = 2$, $nb = nb^* = 1$. Further take $A^*(q^{-1}) = (1-\alpha q^{-1})^2$, $D(q^{-1}) = (1+\alpha q^{-1})^2$, $u(t) = (1-\alpha q^{-1})^2(1+\alpha q^{-1})^2 w(t)$, $w(t)$ white noise of unit variance.

Then the matrix P becomes

$$P(D/K,A^*,u,3,3) = \begin{bmatrix} 1-4\alpha^2+\alpha^4 & -2\alpha(1-\alpha^2) & \alpha^2 \\ 2\alpha(1-\alpha^2) & 1-4\alpha^2+\alpha^4 & -2\alpha(1-\alpha^2) \\ \alpha^2 & 2\alpha(1-\alpha^2) & 1-4\alpha^2+\alpha^4 \end{bmatrix}$$

This matrix, and hence also R, see (4.14), becomes singular if $\alpha = \sqrt{(3-\sqrt{5})/2} \approx 0.618$.

For the IV variant (3.14) which corresponds to $K(q^{-1}) \equiv D(q^{-1}) \equiv 1$, $C(q^{-1}) \equiv q^{-nb}$ in (4.12), (4.13) we can take $A^*(q^{-1})$ and u(t) as above. Then the matrix P becomes

$$P(1,A^*,u,3,3) = \begin{bmatrix} 1-2\alpha^4 & -4\alpha^3 & -2\alpha^2+\alpha^6 \\ 2\alpha & 1-2\alpha^4 & -4\alpha^3 \\ \alpha^2 & 2\alpha & 1-2\alpha^4 \end{bmatrix}$$

It is singular for $\alpha = 0.8922$.

As a numerical illustration for the latter case, 20 independent realizations each containing 500 data were generated. The IV variant (3.14) was applied. The results are given in Table 4.2. They demonstrate indeed the bad accuracy of the estimates.

Parameter	a_1	a_2	b
True value	-1.78	0.80	1.0
Mean of 20 realizations	-19.2	2.60	12.0
Standard dev. of 20 realizations	76.0	8.29	48.7

Table 4.2 Summarized simulation results for the IV variant (3.14). ($\alpha = 0.8922$).

As a comparison the same type of experiment was performed but with $\alpha = 0.992$. Then much more reasonable results were obtained as shown in Table 4.3.

Parameter	a_1	a_2	b
True value	-1.98	0.98	1.0
Mean of 20 realizations	-1.84	0.98	0.86
Standard dev. of 20 realizations	0.49	0.13	0.49

Table 4.3 Summarized simulation results for the IV variant (3.14), ($\alpha = 0.992$).

■

4.3 ANALYSIS OF IV VARIANTS FOR MULTIVARIABLE SYSTEMS

We start this section by investigating the consistency properties of some IV variants applied to MISO systems. MISO model structures were considered in Examples 2.2, 2.3 while a class of extended IV variants was given in Example 3.5. The analysis will to a certain degree parallel that of SISO systems in Section 4.2. In the analysis we will use generalized Sylvester matrices associated with a scalar polynomial and a vector one.

Consider the polynomials

$$\begin{aligned} A(z) &= A_0 z^{na} + \ldots + A_{na} \\ B(z) &= b_0 z^{nb} + \ldots + b_{nb} \end{aligned} \qquad (4.15)$$

the coefficients A_i being $1|m$ vectors. The generalized Sylvester matrix of dimension $(m \cdot \overline{na} + \overline{nb})|m(\overline{na} + \overline{nb})$ where

$$\min(\overline{na} - na,\ \overline{nb} - nb) = 0 \qquad (4.16)$$

is defined as

$$\bar{S}(A,B) = \left[\begin{array}{cccccc} A_0 & A_1 \ldots & & A_{na} & & 0 \\ & \ddots & & & \ddots & \\ 0 & & A_0 & A_1 \ldots & & A_{na} \\ \hline b_0 I & b_1 I \ldots & & b_{nb} I & & 0 \\ & \ddots & & & \ddots & \\ 0 & & b_0 I & b_1 I \ldots & & b_{nb} I \end{array}\right] \begin{array}{l} \overline{nb} \text{ rows} \\ \\ \\ m \cdot \overline{na} \text{ rows} \end{array} \qquad (4.17)$$

The identity matrices in (4.17) have dimension $m|m$.

The rank properties of the rectangular matrix (4.17) are given in Lemma A3.2.

We are now prepared to analyse the IV variant given in Example 3.5 when applied to a MISO system.

Theorem 4.3 Consider the MISO model (2.5) and an extended IV variant without prefiltering and with

$$z(t) = z^*(t) \triangleq K(q^{-1})[u^T(t-1) \ldots u^T(t-na-nb)]^T \qquad (4.18)$$

where $K(q^{-1})$ is a stable and invertible scalar filter.

Suppose that the Assumptions

A2, A4, A5 (existence of an asymptotically stable system corresponding to a unique θ^*)

A3, A6, A7 ($u(\cdot)$, $v(\cdot)$ stationary and independent)

all apply.

Then the consistency condition (4.6) is fulfilled if either of the following two conditions is true

I) $u(t)$ is zero mean white noise with a positive definite covariance matrix

II) $A^*(q^{-1})K(q^{-1})$ is strictly positive real.

Proof. Introduce the notation

$$\psi(t) = [u^T(t-1)\dots\ u^T(t-na-nb)]^T$$

Note that Assumption A5 implies that $n^* = \min(na-na^*, nb-nb^*) = 0$ and that $A^*(z)$, $B^*(z)$ are coprime.

Then R can be written as

$$R = EK(q^{-1})\psi(t)\tilde{\phi}^T(t)$$

$$= EK(q^{-1})\psi(t)\cdot \frac{1}{A^*(q^{-1})} \cdot\psi^T(t)\bar{S}^T(-B^*,A^*)$$

Here the generalized Sylvester matrix $\bar{S}^T(-B^*,A^*)$ has dimension $(na+nb)\cdot nu|(na+nb\cdot nu)$. Due to the assumptions, it now follows from Lemma A3.2 that $\bar{S}^T(-B^*,A^*)$ has rank $na+nb\cdot nu$. It is therefore sufficient to prove that the square matrix (of dimension $(na+nb)\cdot nu|(na+nb)\cdot nu$)

$$P = EK(q^{-1})\psi(t)\cdot \frac{1}{A^*(q^{-1})}\ \psi^T(t)$$

is nonsingular.

Consider first case I). Then P can be written as

$$P = \frac{1}{2\pi i} \oint K(z) \begin{bmatrix} zI \\ \vdots \\ z^{na+nb}I \end{bmatrix} S[z^{-1}I \ldots z^{-(na+nb)}I] \frac{1}{A^*(z^{-1})} \frac{dz}{z} = Q \otimes S$$

where $\otimes$ denotes Kronecker product, $S = Eu(t)u^T(t)$ is positive definite by assumption, and Q is an $(na+nb)|(na+nb)$ matrix with entries

$$Q_{jk} = \frac{1}{2\pi i} \oint \frac{K(z)}{A^*(z^{-1})} z^{j-k} \frac{dz}{z}$$

However, it follows readily from Lemma A3.8 that Q is nonsingular. Thus P must be nonsingular.

Next consider case II). The result follows easily from the above calculations and Lemma A3.3. ∎

Similar to Theorem 4.2 the consistency condition (4.6) is *generically true* for IV variants of the type (4.18) under much weaker conditions than those of Theorem 4.3. Consider for example the instruments

$$z(t) = [K_1(q^{-1})u^T(t-1) \ldots K_{na+nb}(q^{-1})u^T(t-na-nb)]^T$$

where the scalar filters $K_1(z), \ldots, K_{na+nb}(z)$ have all poles and zeros strictly outside the unit circle. Let the input be an ARMA process. Then rank R = nθ is generically true with respect to the parameters of $K_1(z), \ldots, K_{na+nb}(z)$ and the input ARMA parameters. (cf Theorem 4.1 and sufficiency condition I of Theorem 4.3).

We next apply Theorem 4.3 to the special case of a Hammerstein model. Such a model structure was introduced in Example 2.3. A crucial point is then what conditions to impose on the true input $\bar{u}(t)$ in order to make the "auxiliary input" vector

$$u(t) = [\bar{u}(t)\ \bar{u}^2(t) \ldots \bar{u}^m(t)]^T$$

persistently exciting. This aspect is discussed in Appendix A1. In particular it is proved in Lemma A1.1 that if $\bar{u}(t)$ is an ARMA process, then u(t) is p.e. of any order.

Example 4.5 Consistency analysis for Hammerstein models

Consider the model structure (2.5) where

$$\phi^T(t) = [-y(t-1)\dots -y(t-na)\ u^T(t-1)\dots u^T(t-nb)]$$

$$u(t) = [\bar{u}(t)\dots \bar{u}^m(t)]^T$$

Let

$$z^*(t) = K(q^{-1})[u^T(t-1)\dots u^T(t-na-nb)]^T \tag{4.19}$$

where $K(q^{-1})$ is a scalar filter with all poles and zeros outside the unit circle. Such instruments were considered in Example 3.6.

Assume first that $u(t)$ is persistently exciting of order na+nb and that $A^*(z)K(z)$ is strictly positive real. Let the IV variant be given by $z(t) = z^*(t)$. Assume further that the conditions of Theorem 4.3 are fulfilled. It then follows just as in Theorem 4.3 that the consistency condition (4.6) is fulfilled.

Next let $\bar{u}(t)$ be white noise and consider the IV variant with

$$z(t) = z^*(t) - Ez^*(t) \tag{4.20}$$

Introduce the notations

$$m_u = Eu(t) \qquad S = E[u(t)-m_u][u(t)-m_u]^T$$

We then get (see the proof of Theorem 4.3)

$$P = E\{K(q^{-1}) \begin{bmatrix} u(t-1)-m_u \\ \vdots \\ u(t-na-nb)-m_u \end{bmatrix} \cdot \frac{1}{A^*(q^{-1})} [u^T(t-1)\dots u^T(t-na-nb)]\}$$

$$= E\{K(q^{-1}) \begin{bmatrix} u(t-1)-m_u \\ \vdots \\ u(t-na-nb)-m_u \end{bmatrix} \cdot \frac{1}{A^*(q^{-1})} [u^T(t-1)-m_u^T\dots u^T(t-na-nb)-m_u^T]\}$$

Since $\bar{u}(t)$ is assumed to be white noise we have

$$E[u(t-i)-m_u][u^T(t-j)-m_u^T] = S\delta_{ij}$$

and therefore, precisely as in the proof of Theorem 4.3

$$P = Q \otimes S$$

with Q being previously defined in the proof of Theorem 4.3. Clearly Q and S are nonsingular, and we can then conclude that the consistency condition (4.6) is fulfilled. ∎

We will end this section by analysing an IV variant for the full polynomial form of a multivariable system. This model structure was introduced in Example 2.5 and further analysed in Lemma 2.1.

Theorem 4.4. Consider the model structure (2.8) and an IV method given by

$$Z(t) = I \otimes z(t) \tag{4.21a}$$

$$z(t) = [u^T(t-1) \dots u^T(t-nb-m)]^T \tag{4.21b}$$

m being a positive integer, and let $F(q^{-1}) \equiv I$.

Suppose that the Assumptions

A2, A4, A5 (existence of an asymptotically stable system corresponding to a unique θ^*)

A3, A6, A7 ($u(\cdot)$, $v(\cdot)$ stationary and independent)

all apply. Assume further that

i) $m \geq \mu \triangleq$ the controllability index[1] of $G^*(z) = z^{nb-na}G(z^{-1})$

ii) u(t) is white noise with zero mean and nonsingular covariance matrix.

Then the consistency condition (4.6) is fulfilled.

Proof. First note that A5 implies, see Lemma 2.1,

$$\nu = na, \ \delta = \nu \cdot ny$$

where ν = the observability index[1], and δ = the minimal degree[1] of $G^*(z)$.

---

1) We recall that these concepts are reviewed in Appendix 6.

Let $\{G_k\}$ be the Markov parameters of $G^*(z)$, i.e.

$$G^*(z) = \sum_{k=-\infty}^{\infty} G_k z^{-k}$$

We have for convenience written the above sum from $-\infty$ to ∞ but it is clear that $G_k = 0$, $k \le na-nb$ (due to the assumption that $G(z^{-1})$ is strictly proper). Since $u(t)$ is white noise we can write

$$Eu(t-nb-i)x^T(t-j) = Eu(t-nb-i)\circ[G(q^{-1})u(t-j)]^T$$

$$= Eu(t-nb-i)\circ[G^*(q)u(t-j+na-nb)]^T$$

$$= \sum_{k=-\infty}^{\infty} Eu(t-na-i)\circ u^T(t-j-k)G_k{}^T = SG^T_{na+i-j}$$

where $S = Eu(t)u^T(t)$. Hence we have (with $na = \nu$)

$$R = \left[\begin{array}{ccc|c} & X & & I_{nb} \otimes S \\ \hline -SG^T_{\nu}\cdots & & -SG^T_1 & \\ \vdots & & & \bigcirc \\ -SG^T_{\nu+m-1}\cdots & & -SG^T_m & \end{array}\right] = \left[\begin{array}{c|c} X & I_{nb} \otimes S \\ \hline -(I_m \otimes S)M & \bigcirc \end{array}\right]$$

where the block denoted "X" has no influence upon rank R and where M is the following block Toeplitz matrix

$$M = \begin{bmatrix} G^T_{\nu}\cdots & G^T_1 \\ & \\ G^T_{\nu+m-1}\cdots & G^T_m \end{bmatrix}$$

Since S is positive definite we have

$$\text{rank } R = nb\circ nu + \text{rank } M$$

The rank of M is however equal to that of the following usual block Hankel matrix

$$M^* = \begin{bmatrix} G_1 \cdots & G_m \\ & \\ G_{\nu} \cdots & G_{\nu+m-1} \end{bmatrix}$$

We have rank $M^* \leq \bar{\delta}$ with equality if $\nu \geq \bar{\nu}$ and $m \geq \bar{\mu}$, see e.g. Anderson and Jury (1976), where $\bar{\delta}$, $\bar{\nu}$ and $\bar{\mu}$ are the minimal degree, the observability index and, respectively, the controllability index of the strictly proper rational matrix

$$\bar{G}(z) = \sum_{k=1}^{\infty} G_k z^{-k} = G^*(z)-G_0-G_{-1}z-\ldots-G_{na+1-nb}z^{nb-na-1} \triangleq G^*(z)-Q(z)$$

(note that $Q(z) \equiv 0$ if $na-nb \geq 0$). Now, it can be shown that $\bar{\delta} = \delta$ and $\bar{\nu} = \nu$. To this purpose let $\bar{A}(z)^{-1}\bar{B}(z)$ be the row proper factorization of $\bar{G}(z)$. Then we have

$$G^*(z) = \bar{A}(z)^{-1}[\bar{A}(z)Q(z)+\bar{B}(z)] \tag{4.22}$$

Since $\bar{A}(z)$ and $\bar{B}(z)$ are left coprime it follows easily that $\bar{A}(z)$ and $\bar{A}(z)\cdot Q(z)+\bar{B}(z)$ are also left coprime polynomials. Indeed let us assume that there exists a (non-unimodular) polynomial matrix $\bar{L}(z)$ such that $\bar{A}(z) = \bar{L}(z)\tilde{A}(z)$ and that $\bar{L}(z)^{-1}[\bar{A}(z)Q(z)+\bar{B}(z)]$ is a polynomial. However, $\bar{L}(z)^{-1}[\bar{A}(z)Q(z)+\bar{B}(z)] = \tilde{A}(z)Q(z) +\bar{L}(z)^{-1}\bar{B}(z)$ will be a polynomial if and only if $\bar{L}(z)^{-1}\bar{B}(z)$ is a polynimal and we have arrived at a contradiction to the assumption that $\bar{A}(z)$ and $\bar{B}(z)$ are left coprime.

It follows from the above discussion that (4.22) is the row proper factorization of $G^*(z)$. The equalities $\bar{\delta} = \delta$, $\bar{\nu} = \nu$ follow then trivially. Similarly we can show that $\bar{\mu} = \mu$. To finish the proof it only remains to notice that for $m \geq \mu$, $\delta = \nu\cdot ny$ and $\nu = na$ we have

$$\text{rank } R = nb\cdot nu+na\cdot ny$$

which is the desired result. ■

The result of Theorem 4.4 can be generalized to the generic case. Consider for example the instruments given by (4.21a) and

$$z(t) = [\bar{u}^T(t-1)\ldots\ \bar{u}^T(t-nb-m)]^T \qquad m \geq \mu$$

$$\bar{u}(t) = K(q^{-1})u(t)$$

where $K(q^{-1})$ and $K^{-1}(q^{-1})$ are asymptotically stable $nu|nu$ dimensional matrix filters. Further let the input $u(t)$ be an ARMA process. Then rank $R = n\theta$ is *generically true* with respect to the parameters of $K(q^{-1})$ and the input ARMA process.

4.4 CONVERGENCE ANALYSIS OF BOOTSTRAP IV VARIANTS

In this section we will examine the convergence properties of the bootstrap estimators (3.40) introduced in Section 3.3. To simplify the analysis we will treat SISO systems only. Special emphasis will be made on the variants given by (3.41)-(3.43), see Example 3.7.

The analysis of the algorithm (3.40) would require a probabilistic setting. In order to avoid this we will let N tend to infinity. Then the sample covariances appearing in the right hand side of (3.40) will tend, under fairly general conditions, to the corresponding theoretical expectations, cf e.g. Söderström (1975). The iterations become

$$\hat{\theta}^{k+1} = [Ez_1(t,\hat{\theta}^k)z_2^T(t,\hat{\theta}^k)]^{-1}[Ez_1(t,\hat{\theta}^k)y(t)] \tag{4.23}$$

We will here examine the behaviour of this algorithm. The results of this analysis will give valuable insight into the convergence properties of (3.40) for a sufficiently large number of data and, of course, into the consistency properties of the estimators (3.40).

One essential point is to determine conditions which guarantee that the inverse appearing in (4.23) exists. Another topic to be investigated concerns the possible limiting points of (4.23) as the number of iterations, k, tends to infinity. If $\bar{\theta}$ is a possible limiting point it is readily seen from (4.23) that it must satisfy

$$Ez_1(t,\bar{\theta})[y(t)-z_2^T(t,\bar{\theta})\bar{\theta}] = 0 \tag{4.24}$$

The convergence of $\hat{\theta}^k$ to $\bar{\theta}$ is, of course, an interesting aspect as well. A simple linearization gives

$$\begin{aligned}\hat{\theta}^{k+1}-\bar{\theta} &= [Ez_1(t,\hat{\theta}^k)z_2^T(t,\hat{\theta}^k)]^{-1}Ez_1(t,\hat{\theta}^k)\{y(t)-z_2^T(t,\hat{\theta}^k)\bar{\theta}\} \\ &= [Ez_1(t,\bar{\theta})z_2^T(t,\bar{\theta})]^{-1}E[\frac{\partial z_1}{\partial\theta}(t,\theta)|_{\theta=\bar{\theta}}\{y(t)-z_2^T(t,\bar{\theta})\bar{\theta}\} \\ &\quad -z_1(t,\bar{\theta})\bar{\theta}^T\frac{\partial z_2}{\partial\theta}(t,\theta)|_{\theta=\bar{\theta}}](\hat{\theta}^k-\bar{\theta})+(\|\hat{\theta}^k-\bar{\theta}\|^2)\end{aligned} \tag{4.25}$$

Thus, $\hat{\theta}^k$ is locally convergent to $\bar{\theta}$ provided the matrix

$$A = [Ez_1(t,\bar{\theta})z_2^T(t,\bar{\theta})]^{-1}E[\frac{\partial z_1}{\partial\theta}(t,\theta)|_{\theta=\bar{\theta}}\{y(t)-z_2^T(t,\bar{\theta})\bar{\theta}\}-z_1(t,\bar{\theta})\bar{\theta}^T\frac{\partial z_2}{\partial\theta}(t,\theta)|_{\theta=\bar{\theta}}] \tag{4.26}$$

has all eigenvalues within the unit circle.

Remark 1. It is of course desirable that θ^* is the unique solution of (4.24) and that also the corresponding eigenvalue requirement on A is met. ∎

Remark 2. The discussion above concerns local convergence only. To examine global convergence properties of the general algorithm (4.23) is far more difficult since no easy-to-use analysis methods are available. For some specific choices of $z_1(t)$ and $z_2(t)$ the analysis becomes feasible, however, see below. ∎

We now proceed to analyse the two specific bootstrap estimators given by (3.40)-(3.43) in Example 3.7. The results are given as two theorems.

Theorem 4.5. Consider the algorithm BE_1 given by (3.41)-(3.42), (3.43a), (4.23).

Suppose the Assumptions

A2, A4, A5 (existence of an asymptotically stable system corresponding to a unique θ^*)

A3, A6, A7 ($u(\cdot)$, $v(\cdot)$ stationary and independent)

all apply. Then the following results are true:

i) The matrix $Ez_1(t,\hat{\theta}^k)z_2^T(t,\hat{\theta}^k)$ will be nonsingular if $\hat{A}^k(z)$ has all zeros outside the unit circle, $\hat{A}^k(z)$, $\hat{B}^k(z)$ are coprime and either of the following two conditions is satisfied

   1. $\hat{A}^k(z)/A^*(z)$ is strictly positive real and u(t) is persistently exciting of order na+nb

   2. u(t) is an AR process of order less than or equal to na-na*+nb

ii) Consider the set of regular points θ which are such that the matrix $Ez_1(t,\theta)z_2^T(t,\theta)$ exists and is nonsingular. In this set the vector θ^* of true parameters is the only converging point.

iii) The algorithm converges in one iteration to θ^* provided the polynomial $\hat{A}^0(z)$

has all zeros outside the unit circle and the matrix $E\tilde{\varphi}(t,\hat{\theta}^0)\tilde{\varphi}^T(t)$ is nonsingular.

Proof. Note that A6 implies that u(t) is persistently exciting of order na+nb and A5 that $n^* = \min(na-na^*, nb-nb^*) = 0$, $A^*(z)$, $B^*(z)$ being coprime. To prove part i) further note that for BE_1

$$Ez_1(t,\hat{\theta}^k)z_2^T(t,\hat{\theta}^k) = E\tilde{\varphi}(t,\hat{\theta}^k)\tilde{\varphi}^T(t) = S(-\hat{B}^k,\hat{A}^k)P(\hat{A}^k,A^*,u,na+nb,na+nb)S^T(-B^*,A^*)$$

Lemma A3.1 implies that the first and the third matrix of the right hand side are nonsingular. The matrix P is also nonsingular under the given assumptions, see Lemma A3.8.

Assume next that $\bar{\theta}$ is a solution of (4.24) such that the corresponding polynomials $\bar{A}(z)$ and $\bar{B}(z)$ are coprime. Introduce the notations

$$h = [h_1 \dots h_m]^T \tag{4.27}$$

$$H(z) = h_1z+\dots+h_mz^m \equiv \bar{A}(z)B^*(z)-A^*(z)\bar{B}(z) \tag{4.28}$$

$$m = \max(na+nb^*,\ na^*+nb) = na+nb-\min(na-na^*,\ nb-nb^*) = na+nb$$

Then (4.24) implies

$$0 = E\tilde{\varphi}(t,\bar{\theta})[\bar{A}(q^{-1})y(t)-\bar{B}(q^{-1})u(t)] = E\tilde{\varphi}(t,\bar{\theta})[\frac{\bar{A}(q^{-1})B^*(q^{-1})-A^*(q^{-1})\bar{B}(q^{-1})}{A^*(q^{-1})}u(t)]$$

$$= S(-\bar{B},\bar{A})P(\bar{A},A^*,u,m,m)h$$

Since, $S(-\bar{B},\bar{A})$ and P are nonsingular by assumption (otherwise $E\tilde{\varphi}(t,\bar{\theta})\tilde{\varphi}^T(t)$ would be singular, cf the proof of part i)), it follows that $h = 0$, i.e. $\bar{\theta} = \theta^*$ is the unique solution.

In order to complete the proof, the matrix A must be calculated. Since in this case $z_2(t,\theta)$ does not depend on θ we easily obtain from (4.26) for $\bar{\theta} = \theta^*$

$$A = [E\tilde{\varphi}(t)\tilde{\varphi}^T(t)]^{-1}[E\,\frac{\partial}{\partial\theta}\,\tilde{\varphi}(t,\theta)\Big|_{\theta=\theta^*}v(t)] = 0$$

Thus, the algorithm corresponding to this choice of the vectors $z_1(t,\cdot)$ and $z_2(t,\cdot)$ is locally convergent to θ^*. The result can be strengthened since without any approximations we have

$$\hat{\theta}^{k+1}-\theta^* = [E\tilde{\varphi}(t,\hat{\theta}^k)\tilde{\varphi}^T(t)]^{-1}E\tilde{\varphi}(t,\hat{\theta}^k)\{y(t)-\varphi^T(t)\theta^*\}$$

$$= [E\tilde{\varphi}(t,\hat{\theta}^k)\tilde{\varphi}^T(t)]^{-1}E\tilde{\varphi}(t,\hat{\theta}^k)v(t) = 0$$

provided the inverse exists. Then, finally part iii) also is proved. ■

Note the strong result in part iii) of Theorem 4.5. In the asymptotic case ($N \to \infty$) the algorithm converges in one iteration under weak conditions. In the light of the consistency analysis of Section 4.2 this should not be surprising. We may also remark that part ii) of Theorem 4.5 can be viewed as a simple consequence of part iii).

Theorem 4.6. Consider the algorithm BE_2 given by (3.41)-(3.42), (3.43b), (4.23).

Suppose that the Assumptions

A2, A4, A5 (existence of an asymptotically stable system corresponding to a unique θ^*)

A3, A6, A7 ($u(\cdot)$, $v(\cdot)$ stationary and independent)

all apply. Then the following results are true:

i) The matrix $Ez_1(t,\hat{\theta}^k)z_2^T(t,\hat{\theta}^k)$ is well defined and nonsingular if and only if $\hat{A}^k(z)$ has all zeros outside the unit circle, $\hat{A}^k(z)$ and $\hat{B}^k(z)$ are relatively prime and $u(t)$ is persistently exciting of order na+nb.

ii) Consider the set of points θ where the matrix $Ez_1(t,\theta)z_2^T(t,\theta)$ is well defined and nonsingular. In this set the vector θ^* of true parameters is the only possible converging point if either of the following two conditions is satisfied

1. $A^*(z)$ is strictly positive real
2. $u(t)$ is an AR process of order less than or equal to nb.

iii) The algorithm converges locally to θ^* provided the matrix $I-Q^{-1}S$ has all eigenvalues inside the unit circle.

where

$$Q = E\tilde{\varphi}(t)\tilde{\varphi}^T(t) \tag{4.29}$$

$$S = E\tilde{\varphi}(t)\cdot \frac{1}{A^*(q^{-1})} \tilde{\varphi}^T(t) \tag{4.30}$$

Proof. The proof parallels that of Theorem 4.5 to a great extent. In this case we have

$$Ez_1(t,\hat{\theta}^k)z_2^T(t,\hat{\theta}^k) = E\tilde{\varphi}(t,\hat{\theta}^k)\tilde{\varphi}^T(t,\hat{\theta}^k) = S(-\hat{B}^k,\hat{A}^k)P(\hat{A}^k,\hat{A}^k,u,na+nb,na+nb)S^T(-\hat{B}^k,\hat{A}^k) \tag{4.31}$$

which apparently is well defined and nonsingular if and only if $\hat{A}^k(z)$ has all zeros outside the unit circle, and $\hat{A}^k(z)$, $\hat{B}^k(z)$ are coprime. (Note that u(t) is persistently exciting of order na+nb due to A6).

Assume now that (4.31) is nonsingular. The limiting points are then given by the solutions of (4.24). Using the notation (4.28) we get after some simple calculations

$$0 = E\tilde{\varphi}(t,\bar{\theta})[y(t)-\tilde{\varphi}^T(t,\bar{\theta})\bar{\theta}] = E\tilde{\varphi}(t,\bar{\theta})[\frac{B^*(q^{-1})}{A^*(q^{-1})}u(t)+\{\bar{A}(q^{-1})-1\}\frac{\bar{B}(q^{-1})}{\bar{A}(q^{-1})}u(t)-\bar{B}(q^{-1})u(t)]$$

$$= E\tilde{\varphi}(t,\bar{\theta})\frac{H(q^{-1})}{A^*(q^{-1})\bar{A}(q^{-1})}u(t) = S(-\bar{B},\bar{A})P(\bar{A},A^*\bar{A},u,na+nb,na+nb)h \tag{4.32}$$

Under the conditions 1. or 2. the matrix P is nonsingular, cf Lemma A3.8. It is then concluded (as for BE_1) that h = 0 or equivalently that $\bar{\theta} = \theta^*$ is the unique solution of (4.32).

Finally we calculate the matrix A for $\bar{\theta} = \theta^*$. Some straightforward calculation gives

$$A = [E\tilde{\varphi}(t)\tilde{\varphi}^T(t)]^{-1}E[\frac{\partial\tilde{\varphi}(t,\theta)}{\partial\theta}\Big|_{\theta=\theta^*}\frac{1}{A^*(q^{-1})}v(t)$$

$$-\tilde{\varphi}(t)\theta^{*T}\frac{\partial\tilde{\varphi}(t,\theta)}{\partial\theta}\Big|_{\theta=\theta^*}] = Q^{-1}E\{\tilde{\varphi}(t)[1-\frac{1}{A^*(q^{-1})}]\tilde{\varphi}^T(t)\} = I-Q^{-1}S$$

which proves part iii). ■

The convergence of the symmetric algorithm BE_2 clearly requires some restrictions on the system, since the eigenvalues of

$$A = I-Q^{-1}S$$

must lie within the unit circle. In the next example we calculate the eigenvalues and show that the local convergence condition in part iii) of Theorem 4.6 can easily be violated.

Example 4.6 Eigenvalues of $A = I - Q^{-1}S$

We will calculate the eigenvalues under the assumption that u(t) is white noise. We have

$$\tilde{\varphi}(t) = S(-B^*,A^*)\bar{\varphi}(t)$$

$$\bar{\varphi}(t) = \frac{1}{A^*(q^{-1})}[u(t-1)\ldots u(t-na-nb)]^T$$

and thus

$$A = I - S^{-T}(-B^*,A^*)[E\bar{\varphi}(t)\bar{\varphi}^T(t)]^{-1}[E\bar{\varphi}(t)\cdot\frac{1}{A^*(q^{-1})}\bar{\varphi}^T(t)]S^T(-B^*,A^*)$$

Hence the matrix A has the same eigenvalue as

$$I - [E\bar{\varphi}(t)\bar{\varphi}^T(t)]^{-1}[E\bar{\varphi}(t)\cdot\frac{1}{A^*(q^{-1})}\bar{\varphi}^T(t)]$$

The eigenvalues can then be found using Lemma A3.5. There are nb eigenvalues in zero. The remaining na ones are equal to $1-1/A^*(\alpha_i)$ where $\{\alpha_i\}_{i=1}^{na}$ are the roots of $z^{na}+a_1^* z^{na-1}+\ldots+a_{na}^* = z^{na}A^*(z^{-1}) = 0$.

If we specialize to na = na* = 1 we find that the nonzero eigenvalue is $-a^{*2}/(1-a^{*2})$. It is within the unit circle if and only if $a^{*2} < 0.5$. ∎

In the following example we present also a numerical (simulated) illustration of the theoretical analysis carried out in this section.

Example 4.7 Simulation of bootstrap algorithms

A first order system

$$y(t)+a^*y(t-1) = 1.0u(t-1)+e(t)+a^*e(t-1)$$

was used to generate N = 500 data pairs with u(t) a zero mean white noise. The signal to noise ratio was S = 0.3 in all simulations. The two BE algorithms (3.40)-(3.43) were applied. The initial value $\hat{\theta}^0$ was taken equal to the true value θ^* since the interest was in *local* convergence properties.

First the case $a^* = 0.35$ was tried. Then, according to Example 4.6 both eigenvalues of A lie inside the unit circle. (More precisely we have $\lambda_1 = 0$ and $\lambda_2 = -0.14$). The numerical results are given in Table 4.4.

k		0	1	2	3
BE_1	$\hat{a}^k$	0.3500	0.3448	0.3448	0.3448
	$\hat{b}^k$	1.0000	1.0287	1.0287	1.0287
BE_2	$\hat{a}^k$	0.3500	0.3477	0.3440	0.3444
	$\hat{b}^k$	1.0000	1.0116	1.0116	1.0116

Table 4.4. Parameter estimates $\hat{\theta}^k$ versus iteration number k for $a^* = 0.35$.

The same experiment was repeated with $a^* = 0.95$. Then A has one eigenvalue outside the unit circle. (To be exact it is located in -9.26). The numerical results are given in Table 4.5. Here BE_2 gives an unstable model at iteration 2.

k		0	1	2	3
BE_1	$\hat{a}^k$	0.9500	0.9536	0.9534	0.9534
	$\hat{b}^k$	1.0000	0.9525	0.9525	0.9525
BE_2	$\hat{a}^k$	0.9500	0.9444	1.0438	-
	$\hat{b}^k$	1.0000	0.9516	0.9517	-

Table 4.5. Parameter estimates $\hat{\theta}^k$ versus iteration number k for $a^* = 0.95$.

The simulation results are well in accordance with the theory. They demonstrate that BE_1 gives convergence in one iteration and that for BE_2 the values of the A*-coefficients can strongly influence the convergence properties. ∎

4.5 REMARKS AND BIBLIOGRAPHICAL NOTES

When performing the consistency analysis one can work with different convergence concepts. We have chosen to use convergence with probability one. Finigan and Rowe (1974) use the wording "strong consistency" for this choice. In the statistical literature the normal choice is the weaker convergence in probability. A discussion on the use of different convergence concepts in connection with IV methods has been given by Ward (1977). We may remark that the difference between the consistency analyses based on different convergence concepts does not lie in the resulting consistency conditions but rather in the technical details of the proofs.

Theorem 4.1 was inspired by the analysis carried out by Finigan and Rowe (1974). They treated only the SISO case with the instruments (3.11), (3.12), $K(q^{-1}) \equiv 1$. Moreover, the proof given here is much shorter.

Theorem 4.2 is a generalization of the results given previously by Söderström (1974), Söderström and Stoica (1981 a).

For the basic IV method given by (3.15) the consistency condition (3.8) is fulfilled if in addition to the general assumptions we have $H(q^{-1}) \equiv A(q^{-1})$ (the disturbance is white measurement noise), $k \geq na$, and $u(t)$ is either a periodic signal of period k or a white process. For a proof of sufficiency, see Söderström and Stoica (1981 a).

Even though these sufficient conditions are quite restrictive they can according to Theorem 4.1 be used to prove generic consistency under much weaker requirements.

Two other IV variants for SISO systems were mentioned in Section 3.4. Sufficient consistency conditions for these IV variants are given in Söderström and Stoica (1981 a).

Other counterexamples to the general convergence of IV variants than those given in Example 4.4 can be found in Söderström and Stoica (1978).

Theorem 4.3 and Example 4.5 are based on Stoica and Söderström (1982 a, 1982 c), Theorem 4.4 on Stoica and Söderström (1981 d, 1982 c). In Stoica and Söderström (1981 d, 1982 c) we have also given a number of Monte Carlo simulations using the IV variants analysed in Theorem 4.3 and Example 4.5. Finally, Theorems 4.5 and 4.6 are based on Stoica and Söderström (1981 a). In the last reference we also analysed the on-line counterparts of the BE algorithms.

Chapter 5

ACCURACY

5.1 ASYMPTOTIC DISTRIBUTION OF IV ESTIMATORS

In this chapter we will examine the accuracy properties of the general IV estimators described in Chapter 3. The result will be that $\sqrt{N}(\hat{\theta}-\theta^*)$ converges in distribution to a gaussian distribution. We will denote this by

$$\sqrt{N}(\hat{\theta}-\theta^*) \xrightarrow{\text{dist}} N(0,P_{IV})$$

which means that the probability density of the stochastic variable $\sqrt{N}(\hat{\theta}-\theta^*)$ converges, as N tends to infinity, to the gaussian distribution with zero mean and covariance matrix P_{IV}. See Chung (1968) for a discussion of the concept "convergence in distribution". We then may say that asymptotically the estimate $\hat{\theta}$ is gaussian distributed with mean θ^* and covariance matrix P_{IV}/N. Explicit expressions for the covariance matrices will be given as well.

The main tool in the analysis will be a variant of the central limit theorem given by Ljung (1977 a), see Lemma A4.1. In this section we will treat the general extended IV algorithm, while the accuracy properties of the iterative (bootstrap) algorithms are examined in the next section.

For the extended IV estimator we have the following result.

Theorem 5.1. Consider the extended IV method given by (3.23). Suppose that the Assumptions

A2, A4, A5 (existence of an asymptotically stable system corresponding to a unique θ^*)

A3, A6, A7 ($u(\cdot)$, $v(\cdot)$ stationary and independent)

A8 ($Z(\cdot)$ independent of $v(\cdot)$)

all apply. Assume that the estimate $\hat{\theta}$ converges to θ^*. Then $\hat{\theta}$ is asymptotically gaussian distributed,

$$\sqrt{N}\,(\hat{\theta}-\theta^*) \xrightarrow{\text{dist}} N(0,P_{IV}) \tag{5.1}$$

with the covariance matrix P_{IV} given by

$$P_{IV} = (R^TQR)^{-1}R^TQ\cdot E\{[\sum_{i=0}^{\infty} Z(t+i)K_i]\Lambda[\sum_{j=0}^{\infty} K_j^TZ^T(t+j)]\}\cdot QR(R^TQR)^{-1} \tag{5.2}$$

In (5.2), $R = EZ(t)\cdot F(q^{-1})\tilde{\phi}^T(t)$, see (4.5). Further $\{K_i\}_{i=0}^{\infty}$ are defined by

$$\sum_{i=0}^{\infty} K_i z^i = F(z)H(z) \tag{5.3}$$

Proof. For N large enough we clearly can write, cf (3.25)

$$\sqrt{N}\,(\hat{\theta}-\theta^*) = (R^TQR)^{-1}R^TQ[\frac{1}{\sqrt{N}}\sum_{t=1}^{N} Z(t)\cdot F(q^{-1})v(t)]\{1+o(1)\}$$

According to Lemma A4.1 the random vector

$$\frac{1}{\sqrt{N}}\sum_{t=1}^{N} Z(t)\cdot F(q^{-1})v(t)$$

is asymptotically gaussian distributed with zero mean and covariance matrix

$$P = \lim_{N\to\infty}\frac{1}{N}\sum_{t=1}^{N}\sum_{s=1}^{N} EZ(t)[F(q^{-1})v(t)][F(q^{-1})v(s)]^TZ^T(s)$$

It follows then from Lemma A4.2 and its corollary that

$$\sqrt{N}\,(\hat{\theta}-\theta^*) \xrightarrow{\text{dist}} N(0,P_{IV})$$

with

$$P_{IV} = (R^TQR)^{-1}R^TQPQR(R^TQR)^{-1}$$

It thus remains to evaluate the covariance matrix P. Define

$$R_v^F(\tau) \triangleq E[F(q^{-1})v(t)][F(q^{-1})v(t+\tau)]^T$$

We then get

$$P = \lim_{N\to\infty} \frac{1}{N} \sum_{\tau=-N}^{N} (N-|\tau|) EZ(t)R_V^F(\tau)Z^T(t+\tau) = \sum_{\tau=-\infty}^{\infty} EZ(t)R_V^F(\tau)Z^T(t+\tau)$$

$$-\lim_{N\to\infty} \frac{1}{N} \sum_{\tau=-N}^{N} |\tau| EZ(t)R_V^F(\tau)Z^T(t+\tau) \tag{5.4}$$

The first sum converges due to the assumption of stationarity. When examining the second term note first that the assumptions on stationarity imply

$$\| EZ(t)R_V(\tau)Z^T(t+\tau) \| \le C\alpha^{|\tau|}$$

for all τ and some $0 < C < \infty$, $0 < \alpha < 1$. The magnitude of the second term can be evaluated using the following calculations. Let

$$\rho_\tau = EZ(t)R_V^F(\tau)Z^T(t+\tau)$$

Then

$$\| \frac{1}{N} \sum_{\tau=-N}^{N} |\tau| \rho_\tau \| \le \frac{1}{N} \sum_{\tau=-N}^{N} |\tau| \cdot \| \rho_\tau \| \le \frac{1}{N} \sum_{\tau=-N}^{N} |\tau| C\alpha^{|\tau|} \le \frac{2C}{N} \sum_{\tau=0}^{\infty} |\tau| \alpha^{|\tau|} \to 0,\ N \to \infty$$

Therefore

$$\| \lim_{N\to\infty} \frac{1}{N} \sum_{\tau=-N}^{N} |\tau| \rho_\tau \| \le \lim_{N\to\infty} \frac{1}{N} \sum_{\tau=-N}^{N} |\tau| \cdot \| \rho_\tau \| = 0$$

Thus the second term of (5.4) will vanish.

Introduce for convenience $K_i = 0$, $i < 0$. Then we can write

$$P = \sum_{\tau=-\infty}^{\infty} EZ(t)\{ \sum_{i=-\infty}^{\infty} \sum_{j=-\infty}^{\infty} K_i [Ee(t-i)e^T(t+\tau-j)] K_j^T \} Z^T(t+\tau)$$

$$= \sum_{\tau=-\infty}^{\infty} \sum_{i=-\infty}^{\infty} EZ(t)K_i \Lambda K_{\tau+i}^T Z^T(t+\tau) = E[\sum_{i=-\infty}^{\infty} Z(t+i)K_i]\Lambda[\sum_{j=-\infty}^{\infty} Z(t+j)K_j]^T$$

which completes the proof. ∎

Corollary 5.1. If nz = nθ the covariance matrix P_{IV} simplifies to

$$P_{IV} = R^{-1}E\{[\sum_{i=0}^{\infty} Z(t+i)K_i]\Lambda[\sum_{j=0}^{\infty} Z(t+j)K_j]^T\}R^{-T} \tag{5.5}$$

Proof. The result follows directly from (5.2) since R and Q are nonsingular matrices. ∎

Remark. For single output systems the covariance matrix can be written in a slightly different form. Then K_i and Λ are scalars. Due to the assumption of stationarity we have

$$Ez(t+i)z^T(t+j) = Ez(t-j)z^T(t-i)$$

so that the middle matrix of (5.2) can be rewritten as

$$E\{[\sum_{i=0}^{\infty} z(t+i)K_i]\lambda[\sum_{j=0}^{\infty} z(t+j)K_j]^T\} = E\{\sum_{i=0}^{\infty}\sum_{j=0}^{\infty} K_iK_j\lambda z(t-i)z^T(t-j)\}$$

$$= \lambda E[\sum_{i=0}^{\infty} K_i z(t-i)][\sum_{j=0}^{\infty} K_j z^T(t-j)] = \lambda E[K(q^{-1})z(t)][K(q^{-1})z^T(t)] \tag{5.6}$$

∎

In the following example we illustrate the theorem with simulations. Doing so, we also check if the asymptotic results apply reasonably well for data series of finite length.

Example 5.1 *Comparison between sample and asymptotic covariance matrices of IV estimates*

We simulated the following two systems

$$S_1: \quad (1-0.5q^{-1})y(t) = 1.0u(t-1)+(1+0.5q^{-1})e(t)$$

$$S_2: \quad (1-1.5q^{-1}+0.7q^{-2})y(t) = (1.0q^{-1}+0.5q^{-2})u(t)+(1-1.0q^{-1}+0.2q^{-2})e(t)$$

generating for each system 200 realizations, each of length 600. The input u(t) and e(t) were in all cases mutually independent white noises with zero means and unit variances.

The systems were identified in the natural model structure (2.4a) taking na = nb = 1 for S_1 and na = nb = 2 for S_2.

For both systems two IV variants were tried, namely

J_1: $z(t) = [u(t-1)\dots u(t-na-nb)]^T$ (see (3.14))

and

J_2: $z(t) = \frac{1}{A^*(q^{-1})}[u(t-1)\dots u(t-na-nb)]^T$ (see (3.13c))

From the estimates obtained for the 200 different realizations the sample mean and the sample normalized covariance matrix were evaluated as

$$\bar{\theta} = \frac{1}{m}\sum_{i=1}^{m} \hat{\theta}^i$$

$$\hat{P} = \frac{N}{m}\sum_{i=1}^{m} (\hat{\theta}^i-\bar{\theta})(\hat{\theta}^i-\bar{\theta})^T$$

where $\hat{\theta}^i$ denotes the estimate obtained for realization i. N = 600 is the number of data in each realization and m = 200 is the number of realizations. When m tends to infinity we expect that for N sufficiently large $\bar{\theta} \to \theta^*$, $\hat{P} \to P_{IV}$. The deviations from the expected limits for a finite value of m can be approximately evaluated as follows. Assume that N is so large that the asymptotic results of Theorem 5.1 apply. Since the different realizations are independent we get

$$\sqrt{N}(\bar{\theta}-\theta^*) \xrightarrow{\text{dist}} N(0,P_{IV}/m)$$

Thus with 95 percent probability

$$|\bar{\theta}_j-\theta_j^*| \le 1.96\sqrt{\frac{(P_{IV})_{jj}}{mN}}$$

The right hand side is of magnitude 0.01 in the present example.

To evaluate the discrepancy $\hat{P}-P_{IV}$ we substitute $\bar{\theta}$ by θ^* in the expression of $\hat{P}$. This change has only a second order effect. Consider an arbitrary element of $\hat{P} - P_{IV}$. Straightforward calculations give

$$E(\hat{P}_{jk}-P_{IVjk})^2 = \frac{N^2}{m^2} E[\sum_{i_1=1}^{m} (\hat{\theta}_j^{i_1}-\theta_j^*)(\hat{\theta}_k^{i_1}-\theta_k^*) \cdot \sum_{i_2=1}^{m} (\hat{\theta}_j^{i_2}-\theta_j^*)(\hat{\theta}_k^{i_2}-\theta_k^*)]-P_{IVjk}^2$$

$$= \frac{1}{m^2} [(m^2-m)P_{IVjk}^2+mEN^2(\hat{\theta}_j^i-\theta_j^*)^2(\hat{\theta}_k^i-\theta_k^*)^2]-P_{IVjk}^2$$

$$= \frac{1}{m} [EN^2(\hat{\theta}_j^i-\theta_j^*)^2(\hat{\theta}_k^i-\theta_k^*)^2-P_{IVjk}^2]$$

$$= \frac{1}{m} [P_{IVjj}P_{IVkk}+P_{IVjk}^2]$$

The first equality follows from $E\hat{P}_{jk} = P_{IVjk}$. The second equality follows since $\hat{\theta}_j^{i_1}-\theta_j^*$ and $\hat{\theta}_k^{i_2}-\theta_k^*$ are independent if $i_1 \neq i_2$. The third equality is trivial. Finally, in the last equality the gaussian distribution of $\hat{\theta}^i$ is used. If m is reasonably large, say more than 20, we can use the central limit theorem according to which $\hat{P}_{jk}$ is asymptotically gaussian distributed. Then with 95 percent probability

$$|\hat{P}_{jk}-P_{IVjk}| \leq 1.96\sqrt{\frac{P_{IVjj}P_{IVkk}+P_{IVjk}^2}{m}}$$

In particular for the diagonal elements this relation can be rewritten as

$$\left|\frac{\hat{P}_{jj}-P_{IVjj}}{P_{IVjj}}\right| \leq \frac{2.67}{\sqrt{m}}$$

In the present example the right hand side has the value 0.196. This means that the relative error in $\hat{P}_{jj}$ should not be larger than 20 percent, with a probability of 0.95.

The numerical results evaluated from the simulations are shown in Table 5.1. They are well in accordance with the theory. This indicates that the asymptotic results of Theorem 5.1 can be applied also for reasonable lengths, say a few hundred, of the data series.

Distribution parameters: means and normalized covariances	System S_1				System S_2			
	Variant J_1		Variant J_2		Variant J_1		Variant J_2	
	Asympt. expect. values	Sample estim. values	Asympt. expect. values	Sample estim. values	Asympt. expect. values	Sample estim. values	Asympt. expect. values	Sample estim. values
$E\hat{a}_1$	-0.50	-0.50	-0.50	-0.50	-1.50	-1.51	-1.50	-1.50
$E\hat{a}_2$	-	-	-	-	0.70	0.71	0.70	0.70
$E\hat{b}_1$	1.00	1.00	1.00	1.00	1.00	1.00	1.00	1.00
$E\hat{b}_2$	-	-	-	-	0.50	0.49	0.50	0.50
P_{11}	1.25	1.26	1.31	1.37	5.19	6.29	0.25	0.27
P_{12}	-0.50	-0.63	-0.38	-0.56	-7.27	-8.72	-0.22	-0.23
P_{13}	-	-	-	-	-0.24	-0.98	-0.08	0.02
P_{14}	-	-	-	-	6.72	6.74	0.71	0.65
P_{22}	1.25	1.25	1.25	1.23	10.38	12.27	0.20	0.22
P_{23}	-	-	-	-	0.27	1.46	0.06	-0.01
P_{24}	-	-	-	-	-9.13	-9.08	-0.59	-0.54
P_{33}	-	-	-	-	2.04	2.36	2.04	2.15
P_{34}	-	-	-	-	-1.44	-1.97	-1.28	-1.03
P_{44}	-	-	-	-	10.29	8.74	3.21	2.76

Table 5.1. Comparison between asymptotical and sample behaviour of two IV estimates. The sample behaviour shown is estimated from 200 realizations of 600 data pairs each. ■

Theorem 5.1 provides a mean to analyse the accuracy properties of various IV variants. In the following example we show that the covariance matrices of the different IV variants can differ significantly.

Example 5.2 *Comparison of accuracy of different IV variants*

Consider the IV variant (3.11), (3.12) with $K(q^{-1}) = 1$, nc = nb, nd = na. This means that the instruments are given by (neglecting a nonsingular linear transformation as discussed in Example 3.1, see (3.13))

$$z^*(t) = \frac{1}{D(q^{-1})} [u(t-1)\dots u(t-na-nb)]^T$$

We will to some extent examine the influence of the polynomial D on the asymptotic covariance matrix P_{IV}, (5.2).

Assume that the general assumptions are satisfied. Introduce

$$\varphi_0(t) = \frac{1}{A^*(q^{-1})} [u(t-1)\dots u(t-na-nb)]^T$$

Then $\tilde{\varphi}(t) = S(-B^*, A^*)\varphi_0(t)$ where $S(-B^*, A^*)$ is the Sylvester matrix associated with $-B^*$ and A^*. When $H(q^{-1}) = 1$ a simple calculation gives $(F(q^{-1}) \equiv 1)$

$$P_{IV} = \lambda S^{-T}(-B^*, A^*)[Ez^*(t)\varphi_0^T(t)]^{-1}[Ez^*(t)z^{*T}(t)][E\varphi_0(t)z^{*T}(t)]^{-1}S^{-1}(-B^*, A^*)$$

By applying Lemma A3.9 we get

$$P_{IV} \geq \lambda S^{-T}(-B^*, A^*)[E\varphi_0(t)\varphi_0^T(t)]^{-1}S^{-1}(-B^*, A^*)$$

where we have equality for $z^*(t) \equiv \varphi_0(t)$ or equivalently for $D(z) \equiv A^*(z)$. This means that when $H(q^{-1}) \equiv 1$ it is optimal in a general sense (since the whole P_{IV} matrix is optimized) to take $D(z) \equiv A^*(z)$. Note that when $H(q^{-1}) \equiv 1$ the least squares method will in fact give consistent estimates, cf (3.4). It will then have an asymptotic covariance matrix which is "smaller" than that of the IV estimate with $D = A^*$, since we have $P_{IV}(D = A^*) = \lambda[E\tilde{\varphi}(t)\tilde{\varphi}^T(t)]^{-1} \geq \lambda[E\varphi(t)\varphi^T(t)]^{-1} = P_{LS}$.

When $H(z) \neq 1$ the previously discussed instruments, viz. $z^*(t) = \varphi_0(t)$, may no longer be the best choice. Consider for illustration the following first order system

$$y(t)+a^*y(t-1) = b^*u(t-1)+e(t)+ce(t-1)$$

Assume that u(t) and e(t) are mutually independent white noises with zero means and variances σ and λ respectively, and that the IV variant $z^*(t)$ is used to estimate a^* and b^*. Straightforward but somewhat tedious calculations give

$$P_{IV} = \frac{\lambda}{b^{*2}\sigma(1-d^2)} \begin{bmatrix} (1-a^*d)^2(1+c^2-2cd) & -b^*c(1-a^*d)(1-d^2) \\ -b^*c(1-a^*d)(1-d^2) & b^{*2}(1-d^2)(1+c^2) \end{bmatrix}$$

$$\det(P_{IV}) = \left(\frac{\lambda}{\sigma}\right)^2 \frac{(1-a^*d)^2}{b^{*2}(1-d^2)} [1+c^2+c^4-2cd(1+c^2)+c^2d^2]$$

From this expression a simple calculation gives

$$\frac{\partial}{\partial d}\det(P_{IV})_{|d=a^*} = (\frac{\lambda}{\sigma})^2 \frac{1}{b^{*2}} \{\frac{-2a^*(1-a^*d)(1-d^2)+2d(1-a^*d)^2}{(1-d^2)^2}$$

$$\cdot[1+c^2+c^4-2cd(1+c^2)+c^2d^2]+(1-a^{*2})[-2c(1+c^2)+2c^2a^*]\}_{|d=a^*}$$

$$= -2c(\frac{\lambda}{\sigma})^2 \frac{1-d^2}{b^{*2}}(1+c^2-a^*c) \neq 0, \quad \text{if } c \neq 0$$

which shows that the choice $d = a^*$ will in this case not minimize $\det(P_{IV})$ when $c \neq 0$ (i.e. $H(z) \neq 1$).

It can also be seen that the 22 element of P_{IV} is independent of d. Thus, if there would exist some $D = \bar{D}$ such that $P_{IV}(\text{with } D)-P_{IV}(\text{with } D = \bar{D}) \geq 0$ for all D, then the 12 element of P_{IV} must be independent of d as well. This is the case only if $b^*c = 0$. This shows that for the case under study it is not possible to optimize the whole matrix by choosing d.

As a further illustration, $\det(P_{IV})$ was evaluated for i) $d = 0$ (corresponding to the IV variant (3.14)), ii) $d = a^*$ (corresponding to the popular choice $D(z) \equiv A^*(z)$, $C(z) \equiv B^*(z)$, $K(z) = 1$ in (3.11), see also the discussion above; this case is in fact exactly the idealized IV variant we discussed in Example 3.2), and iii) $d = d^*$ = the optimal value of d minimizing $\det(P_{IV})$.

For comparison also a prediction error method (PEM), Ljung (1976), was considered using the following model structure

$$y(t)+ay(t-1) = bu(t-1)+\varepsilon(t)+c\varepsilon(t-1)$$

The PEM parameter estimate $\hat{\theta} = [\hat{a}\ \hat{b}\ \hat{c}]^T$ is obtained by minimizing $V_N(\theta) = \frac{1}{N}\sum_1^N \varepsilon^2(t)$. For gaussian distributed disturbances this estimate is statistically efficient, i.e. its covariance matrix equals the Cramér-Rao lower bound, see Caines and Ljung (1976) who also provide an explicit expression for the covariance matrix, see (6.30).

Using this the (normalized) covariance matrix of $[\hat{a}\ \hat{b}\ \hat{c}]^T$ is found to be

$$\tilde{P}_{PEM} = \lambda \begin{bmatrix} \frac{b^{*2}\sigma(1+a^*c)}{(1-a^{*2})(1-a^*c)(1-c^2)} + \frac{\lambda}{1-a^{*2}} & \frac{b^*c\sigma}{(1-a^*c)(1-c^2)} & -\frac{\lambda}{1-a^*c} \\ \frac{b^*c\sigma}{(1-a^*c)(1-c^2)} & \frac{\sigma}{1-c^2} & 0 \\ -\frac{\lambda}{1-a^*c} & 0 & \frac{\lambda}{1-c^2} \end{bmatrix}^{-1}$$

The covariance matrix for $[\hat{a}\ \hat{b}]^T$ can then be evaluated as the upper left part of $\tilde{P}_{PEM}$. Simple calculations give

$$P_{PEM} = \frac{\lambda}{\sigma} \frac{1}{b^{*2}\sigma+\lambda(c-a^*)^2} \begin{bmatrix} \sigma(1-a^{*2})(1-a^*c)^2 & -b^*c\sigma(1-a^{*2})(1-a^*c) \\ -b^*c\sigma(1-a^{*2})(1-a^*c) & b^{*2}\sigma(1-a^{*2}c^2)+\lambda(c-a^*)^2(1-c^2) \end{bmatrix}$$

$$\det(P_{PEM}) = \left(\frac{\lambda}{\sigma}\right)^2 \frac{(1-a^{*2})(1-c^2)(1-a^*c)^2\sigma}{b^{*2}\sigma+(c-a^*)^2\lambda}$$

The determinant of the covariance matrix of $[\hat{a}\ \hat{b}]^T$ was evaluated for comparison. The numerical values are given in Table 5.2.

Parameter/ Variant	Case 1			Case 2
a^*		-0.8		-0.8
b^*		1.0		1.0
c		0.7		-0.8
σ/λ	0.1	1	10	any value
IV d = 0	1.7301	1.7301	1.7301	2.0496
IV d = a*	1.3365	1.3365	1.3365	0.1296
IV d = d*	1.3113	1.3113	1.3113	0.0785
(d*)	(-0.717)	(-0.717)	(-0.717)	(-0.939)
PEM	0.019	0.138	0.365	0.0168

Table 5.2. Comparison of accuracy of some IV variants and the prediction error method. The figures shown are $\det(P)\cdot b^2\sigma^2/\lambda^2$.

It is clear from the table that the value of d can have a most considerable influence on the accuracy. It also follows that for the present examples a PEM gives much superior accuracies over this basic IVM. ■

Theorem 5.1 can also be used to attempt determination of the optimal IV estimates, that is IV estimates giving maximally possible accuracy. This important aspect will be treated in Section 6.3.

5.2 ASYMPTOTIC DISTRIBUTION OF BOOTSTRAP IV ESTIMATORS

In this section we will evaluate the asymptotic distribution of the iterative estimators BE_1 and BE_2, introduced in Example 3.7. Their convergence properties were analysed in Section 4.4. The accuracy result is given as a theorem.

Theorem 5.2. Consider the algorithms BE_1 and BE_2 given by (3.40) - (3.43). Suppose that the Assumptions

A2, A4, A5 (existence of an asymptotically stable system corresponding to a unique θ^*)

A3, A6, A7 ($u(\cdot)$, $v(\cdot)$ stationary and independent)

all apply. Assume that the estimates are consistent (or equivalently that (4.23) is locally convergent to θ^*).

Then the parameter estimate $\hat{\theta} = \lim_{k\to\infty} \hat{\theta}^k$ is asymptotically gaussian distributed,

$$\sqrt{N}(\hat{\theta}-\theta^*) \xrightarrow{\text{dist}} N(0,P_{BE}) \tag{5.7}$$

The covariance matrix P_{BE} is given by

$$P_{BE_1} = \lambda Q^{-1}[EH(q^{-1})\tilde{\varphi}(t)\cdot H(q^{-1})\tilde{\varphi}^T(t)]Q^{-1} \tag{5.8}$$

for BE_1 and

$$P_{BE_2} = \lambda S^{-1}[E\,\frac{H(q^{-1})}{A^*(q^{-1})}\,\tilde{\varphi}(t)\cdot\frac{H(q^{-1})}{A^*(q^{-1})}\,\tilde{\varphi}^T(t)]S^{-T} \tag{5.9}$$

for BE_2.

Here $\lambda = Ee^2(t)$ and (as in (4.29), (4.30))

$$Q = E\tilde{\varphi}(t)\tilde{\varphi}^T(t) \tag{5.10}$$

$$S = E\tilde{\varphi}(t)\cdot\frac{1}{A^*(q^{-1})}\,\tilde{\varphi}^T(t) \tag{5.11}$$

Proof. In the calculations we will drop higher order terms in the same way as we did when proving Theorem 5.1.

Consider first BE_1. For large N we can write

$$\sqrt{N}\,(\hat{\theta}-\theta^*) = [\frac{1}{N}\sum_{t=1}^{N}\tilde{\varphi}(t,\hat{\theta})\varphi^T(t)]^{-1}[\frac{1}{\sqrt{N}}\sum_{t=1}^{N}\tilde{\varphi}(t,\hat{\theta})v(t)]$$

$$\approx Q^{-1}\{\frac{1}{\sqrt{N}}\sum_{t=1}^{N}\tilde{\varphi}(t)v(t)+[\frac{1}{\sqrt{N}}\sum_{t=1}^{N}\frac{\partial\tilde{\varphi}(t,\theta)}{\partial\theta}\Big|_{\theta=\theta^*}\cdot v(t)](\hat{\theta}-\theta^*)\}$$

$$\approx Q^{-1}\frac{1}{\sqrt{N}}\sum_{t=1}^{N}\tilde{\varphi}(t)v(t)$$

The convergence in distribution (5.7), (5.8) now follows precisely as in Theorem 5.1.

Consider next BE_2. Since $\hat{\theta}$ was assumed to be consistent we have (for N large enough)

$$\sqrt{N}\,(\hat{\theta}-\theta^*) = [\frac{1}{N}\sum_{t=1}^{N}\tilde{\varphi}(t,\hat{\theta})\tilde{\varphi}^T(t,\hat{\theta})]^{-1}\{\frac{1}{\sqrt{N}}\sum_{t=1}^{N}\tilde{\varphi}(t,\hat{\theta})[y(t)-\tilde{\varphi}^T(t,\hat{\theta})\theta^*]\}$$

$$\approx Q^{-1}\{\frac{1}{\sqrt{N}}\sum_{t=1}^{N}\tilde{\varphi}(t)\cdot\frac{1}{A^*(q^{-1})}v(t)+[\frac{1}{\sqrt{N}}\sum_{t=1}^{N}\frac{\partial\tilde{\varphi}(t,\theta)}{\partial\theta}\Big|_{\theta=\theta^*}\frac{1}{A^*(q^{-1})}v(t)](\hat{\theta}-\theta^*)$$

$$-[\frac{1}{\sqrt{N}}\sum_{t=1}^{N}\tilde{\varphi}(t)\theta^{*T}\frac{\partial\tilde{\varphi}(t,\theta)}{\partial\theta}\Big|_{\theta=\theta^*}](\hat{\theta}-\theta^*)\}\approx Q^{-1}\{\frac{1}{\sqrt{N}}\sum_{t=1}^{N}\tilde{\varphi}(t)\cdot\frac{1}{A^*(q^{-1})}v(t)$$

$$+\sqrt{N}\,[E\tilde{\varphi}(t)\frac{A^*(q^{-1})-1}{A^*(q^{-1})}\tilde{\varphi}^T(t)](\hat{\theta}-\theta^*)\} = Q^{-1}\frac{1}{\sqrt{N}}\sum_{t=1}^{N}\tilde{\varphi}(t)\cdot\frac{1}{A^*(q^{-1})}v(t)$$

$$+[I-Q^{-1}S]\sqrt{N}\,(\hat{\theta}-\theta^*)$$

or equivalently

$$\sqrt{N}\,(\hat{\theta}-\theta^*)\approx S^{-1}\cdot\frac{1}{\sqrt{N}}\sum_{t=1}^{N}\tilde{\varphi}(t)\cdot\frac{1}{A^*(q^{-1})}v(t)$$

Then (5.7) and (5.9) follow as in Theorem 5.1. ∎

It is interesting to compare the two covariance matrices P_{BE_1}, (5.8) and P_{BE_2}, (5.9). No strong order relations seem to hold as suggested by the following example.

Example 5.3 Comparison of P_{BE_1} and P_{BE_2}

Let the system be of first order

$$y(t)+a^*y(t-1) = b^*u(t-1)+e(t)+ce(t-1)$$

and let u(t) be zero mean white noise with unit variance. Then the matrix P_{BE_1} can be shown to be

$$P_{BE_1} = \begin{bmatrix} \frac{1-a^{*2}}{b^{*2}}(1+c^2-2a^*c) & -\frac{c}{b^*}(1-a^{*2}) \\ -\frac{c}{b^*}(1-a^{*2}) & 1+c^2 \end{bmatrix}$$

Note that it can be obtained from P_{IV} of Example 5.2 if we take $d = a^*$. Further, calculation of P_{BE_2} gives

$$P_{BE_2} = \begin{bmatrix} \frac{1-a^{*2}}{b^{*2}}(1+c^2-2a^*c) & -\frac{c}{b^*}(1-a^{*2}) \\ -\frac{c}{b^*}(1-a^{*2}) & \frac{1+c^2-2a^*c}{1-a^{*2}} \end{bmatrix}$$

We clearly have

$$P_{BE_1}-P_{BE_2} = \begin{bmatrix} 0 & 0 \\ 0 & \alpha \end{bmatrix}$$

$$\alpha = (1+c^2) - \frac{1+c^2-2a^*c}{1-a^{*2}}$$

In particular $c = 0$ gives $\alpha = -\frac{a^{*2}}{1-a^{*2}} < 0$ while $c = a^*$ gives $\alpha = a^{*2} > 0$.

Since α can have arbitrary sign there cannot exist any strong order relation between P_{BE_1} and P_{BE_2}. Therefore it is not possible to say which of the two algorithms that produces the more accurate estimate. Recall, though, that BE_1 has superior convergence properties. It should usually be preferred in practice. ∎

5.3 REMARKS AND BIBLIOGRAPHICAL NOTES

Theorem 5.1 can be slightly generalized. The result is still true if the instruments Z(t) and the disturbance v(s) are independent only for $t \geq s$, see Söderström and Stoica (1979 a). In the given form, Theorem 5.1 is based on Stoica and Söderström (1981 d, 1982 d, 1983).

Theorem 5.2 is based on Stoica and Söderström (1981 a). There, it is also shown that *idealized* bootstrap algorithms (that is, the BE algorithms obtained from (3.43) by replacing θ with θ^* everywhere) can be similarly analysed. It is in fact possible to show that the idealized BE_1 algorithm gives a covariance matrix equal to (5.8). Thus it gives the same asymptotic accuracy as the realizable BE_1 algorithm. However the covariance matrix for the idealized BE_2 algorithm is neither equal nor necessarily smaller than P_{BE_2}, (5.9), see Stoica and Söderström (1981 a).

Chapter 6

OPTIMAL IV ESTIMATORS

6.1 STATEMENT OF OPTIMALITY PROBLEMS

We will devote this chapter to discussions on how to optimize the accuracy. It is of course always desirable to get as accurate estimates as possible provided that the computational load required be not too high. In this first section we will mainly discuss some general aspects on optimality problems.

One very natural way to measure the accuracy of an estimated model is to use the covariance matrix of the parameter estimates. It will turn out that in most cases it is in fact convenient to use the whole matrix as a *multivariable* measure. Then, if two IV variants give the covariance matrices P_1 respectively P_2 we can say that the first method (corresponding to P_1) is the best one if

$$P_2 - P_1 \geq 0 \tag{6.1}$$

(meaning P_2-P_1 nonnegative definite).

However, in some cases it is not possible to optimize the whole matrix. We then have to work with *scalar* measures of accuracy. Such a measure will then be taken as a (scalar) function of the covariance matrix. We give some examples below. For more details see Goodwin and Payne (1973), Gustavsson et al (1977), Ljung and Söderström (1983).

Example 6.1 Examples of scalar accuracy functions f(P).

Some typical examples of functions f(P) expressing the accuracy are

$$f_1(P) = \det(P) \tag{6.2}$$

$$f_2(P) = \operatorname{tr}(WP) \tag{6.3}$$

where W is a symmetric nonnegative definite matrix. One way to choose W is the following. Start with a criterion $V(\theta)$ which is a function of the model parameter θ. A typical case for a SISO system would be

$$V(\theta) = \sum_{i=1}^{\infty} (h_i - h_i^*)^2 = \frac{1}{2\pi i} \oint [\frac{B(z)}{A(z)} - \frac{B^*(z)}{A^*(z)}][\frac{B(z^{-1})}{A(z^{-1})} - \frac{B^*(z^{-1})}{A^*(z^{-1})}] \frac{dz}{z} \tag{6.4}$$

where $\{h_i\}_{i=1}^{\infty}$ are the weighting function coefficients of the model (i.e. $\sum_{i=1}^{\infty} h_i z^i \equiv B(z)/A(z)$) and $\{h_i^*\}$ have a similar meaning for the true system. For the criterion $V(\theta)$ to be a meaningful one, it should be minimized for $\theta = \theta^*$. This is certainly true for the specific case (6.4). Then take as an accuracy measure of the model the expected value of $V(\hat{\theta})$ where expectation is with respect to the uncertainties in $\hat{\theta}$. Since, at least for long data series, we can assume that $\hat{\theta} \approx \theta^*$ we have approximately

$$f(P) = E_{\hat{\theta}} V(\hat{\theta}) \approx E_{\hat{\theta}}[V(\theta^*) + V'(\theta^*)(\hat{\theta}-\theta^*) + \frac{1}{2}(\hat{\theta}-\theta^*)^T V''(\theta^*)(\hat{\theta}-\theta^*)]$$

$$= V(\theta^*) + \frac{1}{2} \operatorname{tr}\{V''(\theta^*) E(\hat{\theta}-\theta^*)(\hat{\theta}-\theta^*)^T\} = V(\theta^*) + \frac{1}{2N} \operatorname{tr}\{V''(\theta^*)P\} \tag{6.5}$$

We thus find that a natural choice of the weighting matrix W in (6.3) would be

$$W = V''(\theta^*) \tag{6.6}$$

Accuracy functions of the types (6.2), (6.3) can also be used to find optimal input signals, see Chapter 7. Let us only mention here that use of (6.2) is often referred as D-optimality while use of (6.3) sometimes is called A-optimality, cf e.g. Mehra (1974).

Note that both the functions $f_1(P)$ and $f_2(P)$ are monotonically increasing. It is easy to see that for P positive definite, ΔP nonnegative definite,

$$f_i(P+\Delta P) \geq f_i(P) \quad i = 1,2$$

and that the equality holds if and only if $\Delta P = 0$. This is a general necessary requirement for any meaningful accuracy function. ■

To formulate an optimization problem we must also specify the independent variables. In the context of our discussion in this chapter we will call them design variables. We will let ν denote the vector of such design variables. Let us examplify.

Example 6.2 Some design variables ν

We now present some typical examples of design vectors ν. These examples will be analysed later on.

Consider *first* a SISO system and the IV variant (3.11)-(3.12) with $K(q^{-1}) \equiv 1$, i.e.

$$z(t) = [-\eta(t-1)\dots -\eta(t-na)\ u(t-1)\dots u(t-nb)]^T \tag{6.7a}$$

where

$$D(q^{-1})\eta(t) = C(q^{-1})u(t) \tag{6.7b}$$

$$C(z) = c_0+c_1z+\dots+c_{nc}z^{nc} \tag{6.7c}$$

$$D(z) = 1+d_1z+\dots+d_{nd}z^{nd} \tag{6.7d}$$

Assume that the degrees nc and nd are given. We can then consider the polynomial coefficients $\{c_i, d_j\}$ as design variables, i.e.

$$\nu = [c_0\dots c_{nc}\ d_1\dots d_{nd}]^T \tag{6.8}$$

As a *second* example consider again a SISO system and assume that a vector of instruments $z(t) = z_0(t)$ in a basic IV method was selected. Then we may add instruments, i.e. augment the instrumental vector to get an extended IV variant. We then have

$$z(t) = \begin{bmatrix} z_0(t) \\ z_1(t) \end{bmatrix} \tag{6.9}$$

where we now can treat the additional instruments $z_1(t)$ as design variables while keeping $z_0(t)$ fixed.

This case covers in itself many possibilities. The vector $z_1(t)$ can be considered as consisting of completely free variables, it can be described with some filter parameters such as (6.8), or it can have a fixed form but variable dimension.

As a *third* example consider the general extended IV variant (3.23). We then can treat the matrix of instruments, the prefilter and the weighting matrix Q as design variables. This can be written symbolically as

$$\nu = \begin{pmatrix} Z(t) \\ F(q^{-1}) \\ Q \end{pmatrix} \tag{6.10}$$

It is of course possible to consider modified problems where parts of ν in (6.10) are given or have a given structure with some free parameters. ∎

We can now formulate the optimization problem as follows

> *Optimization problem*
>
> Find the vector ν of design variables which minimizes the scalar accuracy function $f(P(\nu))$.

The optimal design variables ν_{opt} will in general depend on the true parameter vector θ^* and the noise filter $H(q^{-1})$ for the simple reason that the accuracy function $f(P)$ depends on θ^* and $H(q^{-1})$. This dependence is, of course, in a sense unfortunate since θ^* and $H(q^{-1})$ are not known. (Otherwise there would be no need for identification). However, this is the same dilemma as encountered in some other optimization problems in identification like e.g. optimal input design. Hence there exists some experience in handling it.

A typical way to overcome this obstacle is to use consistent estimates $\hat{\theta}$ and $\hat{H}(q^{-1})$ instead of the true θ^* and $H(q^{-1})$. Some iterations could possibly be made to improve the estimates. An algorithm exploiting this idea will be presented in Section 6.3.

Remark. The dependence of ν_{opt} on the unknown noise properties (i.e. $H(q^{-1})$) can be eliminated, at least in principle, by using a minimax approach. Then the design variables are determined as the solution to

$$\min_{\nu} \max_{\{v(t)\};Ev^2(t)=1} f(P(\nu)) \qquad (6.11)$$

Unfortunately, the optimization problem (6.11) is often intractable, see Stoica and Söderström (1981 b). Wong and Polak (1967) have tried to solve the problem (6.11) in a slightly reformulated form. However, in that form it is difficult to take the necessary constraints into account. The consistency condition (3.28) constitutes such a constraint. This was in fact overlooked in the analysis by Wong and Polak (1967), see Stoica and Söderström (1981 b). Neglecting this constraint and solving the optimization problem from a pure "algebraic" point of view, they arrived at $z(t) = \tilde{\varphi}(t)$ as minimax optimal instruments. Note that these instruments were presented as an idealized IV variant in Example 3.2. They can also be seen as a motive for the construction of the bootstrap algorithms in Example 3.7. We also used them for comparison purposes in Example 5.2. However, these IVs do not seem to be optimal in any well-defined sense. ∎

After optimizing the accuracy function $f(P(\nu))$ with respect to the design variables we get an optimal accuracy that depends on the input (more generally, the experimental condition). If the input can be chosen it could be selected to improve the accuracy further. This topic will be discussed in Chapter 7.

6.2 SOME OPTIMIZATION EXAMPLES

In the foregoing section we defined some different design variables and indicated how they could be used to optimize the accuracy. In particular we gave in Example 6.2 three (out of virtually infinitly many) different possibilities to introduce design variables. In this section we will show by means of some examples that the first two choice of design variables of Example 6.2 will in general not lead to any neat solutions.

Example 6.3 *Optimal filter parameters in a basic IV variant.*

Consider the IV variant given by (6.7) and let the design vector ν be as in (6.8). Assume that nc = nb, nd = na. As noted in Example 3.1 the coefficients $c_o \dots c_{nc}$ will have no influence on the estimate $\hat{\theta}$. Moreover, we analysed the accuracy of this class of IV variants to some extent in Example 5.2. In particular we showed that

- If $H(q^{-1}) \equiv 1$ then the whole covariance matrix is optimized (in the sense of (6.1)) for $D(z) \equiv A^*(z)$. This means that the optimal accuracy is obtained with $z(t) = \tilde{\varphi}(t)$.

- If $H(q^{-1}) \not\equiv 1$ no neat results seem to apply. Indeed, for first order systems, where there is only one design parameter (namely d_1) and for the criterion $f(P) = \det(P)$, (6.2), we saw that there is no analytic solution to the optimization problem. In particular, the optimum is no longer obtained for $D(z) \equiv A^*(z)$. ∎

Example 6.4 Augmentation of the IV vector

Consider the system

$$y(t)+a^*y(t-1) = b^*u(t-1)+e(t)+ce(t-1) \tag{6.12}$$

which is to be identified in the model structure

$$y(t)+ay(t-1) = bu(t-1)+\varepsilon(t)$$

using an extended IV variant with $Q = I$, $F(z) \equiv 1$ (no prefilter) and

$$z(t) = [u(t-1)\dots\ u(t-k)]^T \qquad (k \geq 2)$$

Assume that $u(\cdot)$ and $e(\cdot)$ are mutually independent white noises with zero means and variances σ respectively λ. Then the matrix R (of dimension $k|2$) becomes, see (4.5)

$$R = \sigma \begin{bmatrix} 0 & 1 \\ -b^* & 0 \\ -(-a^*)b^* & \vdots \\ \vdots & \\ -(-a^*)^{k-2}b^* & 0 \end{bmatrix}$$

The asymptotic covariance matrix, P_{IV}, can be evaluated from Theorem 5.1. The result will be

$$P_{IV} = \frac{\lambda}{\sigma}\begin{bmatrix} \frac{b^{*2}}{1-a^{*2}}(1-a^{*2k-2}) & 0 \\ 0 & 1 \end{bmatrix}^{-1} \begin{bmatrix} 0 & -b^* \dots -(-a^*)^{k-2}b^* \\ 1 & 0 \dots\dots\ 0 \end{bmatrix}$$

$$\cdot \begin{bmatrix} 1+c^2 & c & & 0 \\ c & \ddots & \ddots & \\ & \ddots & \ddots & c \\ 0 & & c & 1+c^2 \end{bmatrix} \begin{bmatrix} 0 & 1 \\ -b^* & 0 \\ \vdots & \vdots \\ -(-a^*)^{k-2}b^* & 0 \end{bmatrix} \begin{bmatrix} \frac{b^{*2}}{1-a^{*2}}(1-a^{*2k-2}) & 0 \\ 0 & 1 \end{bmatrix}^{-1}$$

$$= \frac{\lambda}{\sigma}\begin{bmatrix} \frac{1-a^{*2}}{b^{*2}(1-a^{*2k-2})} & 0 \\ 0 & 1 \end{bmatrix} \{(1+c^2)\begin{bmatrix} \frac{b^{*2}(1-a^{*2k-2})}{1-a^{*2}} & 0 \\ 0 & 1 \end{bmatrix}$$

$$+c\begin{bmatrix} \frac{-2a^{*}\, b^{*2}(1-a^{*2k-4})}{1-a^{*2}} & -b^{*} \\ -b^{*} & 0 \end{bmatrix}\}\begin{bmatrix} \frac{1-a^{*2}}{b^{*2}(1-a^{*2k-2})} & 0 \\ 0 & 1 \end{bmatrix}$$

$$= \frac{\lambda}{\sigma}\begin{bmatrix} \frac{1-a^{*2}}{b^{*2}(1-a^{*2k-2})^2}\{(1+c^2)(1-a^{*2k-2})-2a^{*}c(1-a^{*2k-4})\} & -\frac{c}{b^{*}}\frac{1-a^{*2}}{1-a^{*2k-2}} \\ -\frac{c}{b^{*}}\frac{1-a^{*2}}{1-a^{*2k-2}} & 1+c^2 \end{bmatrix} \quad (6.13)$$

Since the 22 element does not depend on k, "strong order relations" like (6.1) cannot hold. If P_k denotes the covariance matrix corresponding to k then $P_{k_1}-P_{k_2}$ is indefinite (unless c = 0, which is a quite particular case). Consider then the 11 elements. We get for some special cases

$$P_2^{(11)} = \frac{\lambda}{b^{*2}\sigma}(1+c^2)$$

$$P_3^{(11)} = \frac{\lambda}{b^{*2}\sigma}\left[\frac{(1+c^2)(1+a^{*2})-2a^{*}c}{(1+a^{*2})^2}\right]$$

$$P_\infty^{(11)} = \frac{\lambda}{b^{*2}\sigma}[(1-a^{*2})(1+c^2-2a^{*}c)]$$

Depending on the values of a* and c it is possible to have an arbitrary sign of the differences

$$P_2^{(11)}-P_3^{(11)}, \quad P_2^{(11)}-P_\infty^{(11)}, \quad P_3^{(11)}-P_\infty^{(11)}$$

As a conclusion we note that for $c \neq 0$ there are *no generally valid rules* for choosing an optimal dimension k of the instrumental vector z(t).

For c = 0 the 12 and 21 elements of P_k are zero and the 11 element becomes

$$p_k^{(11)} = \frac{\lambda}{b^{*2}\sigma} \frac{1-a^{*2}}{1-a^{*2(k-1)}}$$

Hence we clearly have in this case

$$P_{k_1} \geq P_{k_2} \qquad k_1 < k_2 \tag{6.14}$$

which is a neat result. See Stoica and Söderström (1981 b) for its extension to an arbitrary order of the system. Unfortunately, similar to the case analysed in Example 6.3 it holds only for white equation errors and is thus of limited interest.

■

6.3 OPTIMAL IV ESTIMATORS AND APPROXIMATE IMPLEMENTATIONS

Optimal IV estimators

The optimization problem we introduced in Section 6.1 was analysed for some special choices of the design variables in Section 6.2. For the cases examined there no neat analytic solutions exist.

However, in this section we will treat extended IV variants and use the prefilter $\Gamma(q^{-1})$, the weighting matrix Q and the instruments Z(t) as design variables. Then there is a neat analytic solution of the optimization problem, even of the strong form (6.1). To show this we will first derive a lower bound on the asymptotic covariance matrix P_{IV}, (5.2). Then we will show how this lower bound can be attained with appropriate choices of $F(q^{-1})$, Q and Z(t).

Theorem 6.1. The covariance matrix P_{IV}, (5.2) fulfils

$$P_{IV} \geq P_{IV}^{opt} \triangleq \{E[H(q^{-1})^{-1}\tilde{\phi}^T(t)]^T \Lambda^{-1}[H(q^{-1})^{-1}\tilde{\phi}^T(t)]\}^{-1} \tag{6.15}$$

Moreover, equality in (6.15) holds if and only if

$$R^T Q \sum_{j=0}^{\infty} Z(t+j)K_j = M[\Lambda^{-1}H(q^{-1})^{-1}\tilde{\phi}^T(t)]^T \tag{6.16}$$

for some constant and nonsingular matrix M. The matrices $\{K_j\}$ are defined by (5.3).

Proof. Note that P_{IV}^{opt} as defined by (6.15) is positive definite due to the general assumptions. Introduce the notations

$$w(t) = R^T Q Z(t)$$

$$\alpha(t) = \sum_{j=0}^{\infty} w(t+j)K_j$$

$$\beta(t) = [H(q^{-1})^{-1}\tilde{\phi}^T(t)]^T$$

Then we can write

$$R^T QR = Ew(t)\cdot F(q^{-1})\tilde{\phi}^T(t) = Ew(t)\cdot \sum_{j=0}^{\infty} K_j q^{-j} H(q^{-1})^{-1}\tilde{\phi}^T(t)$$

$$= E[\sum_{j=0}^{\infty} w(t+j)K_j][H(q^{-1})^{-1}\tilde{\phi}^T(t)] = E\alpha(t)\beta^T(t)$$

and thus

$$P_{IV} = [E\alpha(t)\beta^T(t)]^{-1}[E\alpha(t)\Lambda\alpha^T(t)][E\beta(t)\alpha^T(t)]^{-1} \tag{6.17}$$

Since Q is assumed to be positive definite and R of full rank it follows easily that the matrix (6.17) is positive definite. The relation (6.15) can then be written as

$$[E\beta(t)\Lambda^{-1}\beta^T(t)]-[E\beta(t)\alpha^T(t)][E\alpha(t)\Lambda\alpha^T(t)]^{-1}[E\alpha(t)\beta^T(t)] \geq 0 \tag{6.18}$$

or equivalently

$$E\begin{bmatrix}\beta(t)\Lambda^{-1/2}\\ \alpha(t)\Lambda^{1/2}\end{bmatrix}[\Lambda^{-1/2}\beta^T(t) \quad \Lambda^{1/2}\alpha^T(t)] \geq 0$$

which is evident. Furthermore, it follows from Lemma A3.9 that equality in (6.18) holds if and only if $\alpha(t) = M\beta(t)\Lambda^{-1}$, where M is an arbitrary non-singular matrix. The theorem is thus proved. ■

We next turn to the question of if (and how) the lower bound P_{IV}^{opt} of the covariance matrix can be attained. We present two particular solutions.

Solution 1

$$Z_1(t) = [\Lambda^{-1}H(q^{-1})^{-1}\tilde{\phi}^T(t)]^T$$

$$F_1(q^{-1}) \equiv H(q^{-1})^{-1} \tag{6.19}$$

$$(Q_1 = I)$$

$$\hat{\theta} = [\sum_{t=1}^{N} Z_1(t)\cdot F_1(q^{-1})\phi^T(t)]^{-1}[\sum_{t=1}^{N} Z_1(t)\cdot F_1(q^{-1})y(t)]$$

Solution 2

$$Z_2(t) = [H(q)^{-T}\Lambda^{-1}H(q^{-1})^{-1}\tilde{\phi}^T(t)]^T$$

$$F_2(q^{-1}) \equiv I \tag{6.20}$$

$$(Q_2 = I)$$

$$\hat{\theta} = [\sum_{t=1}^{N} Z_2(t)\cdot\phi^T(t)]^{-1}[\sum_{t=1}^{N} Z_2(t)\cdot y(t)]$$

Some comments on these solutions are in order.

It is not difficult to see that both (6.19) and (6.20) fulfil (6.16) with $M = R^TQ$, as well as the general assumptions on IV variants.

For both solutions we have $nz = n\theta$. There are thus no reasons to use additional instruments as long as the optimal accuracy is concerned. In particular the parameter estimates are given by (3.29) rather than (3.23). The matrix Q becomes superfluous. Also note that (6.19) defines an idealized extended IV variant as discussed in Example 3.4. In the scalar case ($ny = 1$) the factor Λ^{-1} is (for both solutions) irrelevant and can be omitted.

Some comments on the computation of $Z_2(t)$ must also be made. Since the expression of $Z_2(t)$ involves $H(q)^{-T}$ with argument q (not q^{-1}) some care must be used in the interpretation. One way to compute $Z_2(t)$ is to transform to the frequency domain. Let F denote the (discrete) Fourier transform and let $\phi_v(\omega)$ denote the spectral density matrix of v(t), that is $\phi_v(\omega) = H(e^{-i\omega})\Lambda H^T(e^{i\omega})$. Then $Z_2(t)$ can be computed by using frequency domain techniques as follows

$$Z_2(t) = [F^{-1}\{\phi_v^{-1}(\omega)\cdot F\tilde{\phi}^T(t)\}]^T$$

It is clear that to evaluate $Z_2(t)$ it is not necessary to have a parametric model of the noise. Only the noise spectral density matrix, possibly in a nonparametric form, is needed. For some applications this may be an advantage. More details on this aspect for the SISO case can be found in Engle (1980).

Another possibility is to organize the computations in the time domain and to use both forward and backward filtering. In particular this means that all data are needed to compute $Z_2(t)$ for any t. A recursive algorithm like (3.21) then cannot be used in a true on-line fashion. The number of operations needed to implement (6.20) is though not significantly different from the requirement for (6.19). To illustrate the filtering procedure in (6.20) consider the following example.

Example 6.5 Illustration of backward filtering

Consider a scalar system and let

$$H(q^{-1}) = \frac{C(q^{-1})}{D(q^{-1})} = \frac{1+c_1q^{-1}+\ldots+c_{nc}q^{-nc}}{1+d_1q^{-1}+\ldots+d_{nd}q^{-nd}}$$

To find $Z_2(t)$ first compute

$$\phi^F(t) = \frac{D(q^{-1})}{C(q^{-1})}\tilde{\phi}(t)$$

Then $Z_2(t)$ is formally given by

$$Z_2(t) = \frac{D(q)}{C(q)}\phi^F(t)$$

where the factor $1/\lambda$ is omitted as it has no importance here.

In practice $Z_2(t)$ will be computed recursively through

$$Z_2(t) = -c_1Z_2(t+1)-\ldots-c_{nc}Z_2(t+nc)+\phi^F(t)+d_1\phi^F(t+1)+\ldots+d_{nd}\phi^F(t+nd)$$

for t = N, N-1,... (i.e. *backward* in time). ■

The solutions (6.19) and (6.20) require knowledge of the undisturbed output $x(t)$ as well as of the noise autocorrelation as described with Λ and $H(q^{-1})$. In practice a bootstrapping technique must therefore be used to implement the optimal IV estimate. The basic idea of such an algorithm is to combine in an iterative manner the optimal IV method with a procedure for estimation of the noise parameters. We will present such an implementation later in this section. Such a bootstrap procedure will have nice asymptotical properties. In particular, it will be shown that if in (6.19) or (6.20), $x(t)$ or (equivalently θ^*), Λ and $H(q^{-1})$ are replaced by some consistent estimates, then the IV estimates so obtained have asymptotically the optimal distribution as given by (5.1) with $P_{IV} = P_{IV}^{opt}$.

Comparison with PEM

To complete the analysis, the maximum achievable accuracy in the class of extended IV estimates will be compared with the accuracy obtained when a prediction error method (PEM) is used. The accuracies are expressed with the covariance matrices of the estimated parameters. We shall assume that the system dynamics is parameterized in the *same* way both for the IVM and for the PEM. This natural assumption implies that the PEM is applied in the following model structure

$$M_6:\ y(t) = \phi^T(t)\theta + H(q^{-1},\theta,\beta)\varepsilon(t,\theta,\beta) \tag{6.21}$$

where $\{\varepsilon(t,\theta,\beta)\}$ are the prediction errors, β is a vector of additional (unknown) parameters, and $H(q^{-1},\theta,\beta)$ is a model of the noise shaping filter $H(q^{-1})$, cf (2.11). We shall assume that $H(q^{-1},\theta,\beta)$ is a canonical parameterization. Then the equality $H(z,\theta^*,\beta) \equiv H(z)$ is fulfilled for a unique vector β. Let this vector be denoted by β^*. We will call β^* the vector of true noise parameters. When a PEM is used we will estimate the extended parameter vector

$$\psi = \begin{bmatrix} \theta \\ \\ \beta \end{bmatrix} \tag{6.22}$$

Before reviewing some properties of PEMs we present some examples of the model parameterization (6.21). We will return to these examples several times in this chapter.

Example 6.6 Some model parameterizations

First note that (6.21) can also be written as

$$A(q^{-1},\theta)y(t) = B(q^{-1},\theta)u(t)+H(q^{-1},\theta,\beta)\varepsilon(t,\theta,\beta)$$

cf (2.1), (2.3). Assume first that $H(q^{-1},\theta,\beta)$ can be completely described by using the parameter vector β only. Then if a (left) matrix fraction description (MFD) is used to parameterize $H(q^{-1},\beta)$ we get the model structure

$$M_7: A(q^{-1},\theta)y(t) = B(q^{-1},\theta)u(t)+D(q^{-1},\beta)^{-1}C(q^{-1},\beta)\varepsilon(t,\theta,\beta) \qquad (6.23)$$

where C and D are the following polynomial matrices of dimension ny|ny

$$C(q^{-1},\beta) = I+C_1(\beta)q^{-1}+\ldots+C_{nc}(\beta)q^{-nc}$$

$$D(q^{-1},\beta) = I+D_1(\beta)q^{-1}+\ldots+D_{nd}(\beta)q^{-nd}$$

In the scalar case (ny = 1) the natural choice of β is

$$\beta = [c_1 \ldots c_{nc}\ d_1 \ldots d_{nd}]^T \qquad (6.24)$$

i.e. all the coefficients of $C(\cdot)$ and $D(\cdot)$ are model parameters. Then the model structure (6.23) can also be written as (assuming θ contains all the coefficients of $A(\cdot)$ and $B(\cdot)$)

$$M_7': y(t) = \frac{B(q^{-1})}{A(q^{-1})}u(t)+\frac{C(q^{-1})}{A(q^{-1})D(q^{-1})}\varepsilon(t,\theta,\beta) \qquad (6.25)$$

Note that the polynomial $A(q^{-1})$ appears in both the denominators in (6.25), so that the system dynamics and the noise part of the model have some common pöles. At a first sight this might seem to be a restriction. In practice, however, there will often be some "internal" disturbances which pass at least partially through the system, thus exciting some of the system modes. In such cases the structure (6.25) will be an appropriate one.

Another possibility is to parameterize $H(q^{-1},\theta,\beta)$ as

$$H(q^{-1},\theta,\beta) = A(q^{-1},\theta)\bar{H}(q^{-1},\beta) \qquad (6.26)$$

where $\bar{H}(q^{-1},\beta)$ is a matrix filter depending only on β. This alternative will give the model structure

$$\mathcal{M}_8: y(t) = A(q^{-1},\theta)^{-1}B(q^{-1},\theta)u(t)+\bar{H}(q^{-1},\beta)\varepsilon(t,\theta,\beta) \tag{6.27}$$

Note that now the system dynamics, i.e. the transfer function from u(•) to y(•), and the noise shaping filter are described with independent parameter vectors. In other words, we can say that for (6.27) the system dynamics and the noise model are "decoupled". The filter $\bar{H}(q^{-1},\beta)$ can again be parameterized as a MFD. This will in the scalar case with β as in (6.24) give

$$\mathcal{M}_8': y(t) = \frac{B(q^{-1})}{A(q^{-1})}u(t)+\frac{C(q^{-1})}{D(q^{-1})}\varepsilon(t,\theta,\beta) \tag{6.28}$$

Note that the model structure $\mathcal{M}_7$ can be obtained by appropriately constraining $\mathcal{M}_8$. In such a case $\mathcal{M}_8$ will contain some redundant parameters. The converse is also true.

■

Let us now review some properties of the prediction error estimates, see Ljung (1976), Caines (1976 a) for details. The PE estimate $\hat{\psi}$ of ψ, see (6.21), (6.22) is obtained as the minimizing element of

$$V^N(\psi) = \det\left(\frac{1}{N}\sum_{t=1}^{N}\varepsilon(t,\psi)\varepsilon^T(t,\psi)\right) \tag{6.29}$$

Under the general assumptions given previously $\hat{\psi}$ will be a consistent estimate of $\psi^* = [\theta^{*T}\ \beta^{*T}]^T$. It will also be asymptotically gaussian distributed, Caines and Ljung (1976),

$$\sqrt{N}(\hat{\psi}-\psi^*) \xrightarrow{dist} N(0,P_{PEM})$$

with

$$P_{PEM} = \{E[\frac{\partial\varepsilon(t,\psi)}{\partial\psi}\Big|_{\psi=\psi^*}]^T\Lambda^{-1}[\frac{\partial\varepsilon(t,\psi)}{\partial\psi}\Big|_{\psi=\psi^*}]\}^{-1} \tag{6.30}$$

See Theorem 5.1 for a similar result for IV estimates.

A consistent estimate of Λ is given by

$$\hat{\Lambda} = \frac{1}{N}\sum_{t=1}^{N}\varepsilon(t,\hat{\psi})\varepsilon^T(t,\hat{\psi}) \tag{6.31}$$

To evaluate the covariance matrix P_{PEM} we must find the gradient of the prediction error. For the general model structure (6.21) we obtain

$$\varepsilon(t,\psi) = H(q^{-1},\theta,\beta)^{-1}[y(t)-\phi^T(t)\theta]$$

and after some straightforward calculations

$$\left.\frac{\partial\varepsilon(t,\psi)}{\partial\theta_i}\right|_{\psi=\psi^*} = -H(q^{-1})^{-1}\left.\frac{\partial H(q^{-1},\psi)}{\partial\theta_i}\right|_{\psi=\psi^*} e(t)-H(q^{-1})^{-1}\phi^T(t)e_i \tag{6.32}$$

$$\left.\frac{\partial\varepsilon(t,\psi)}{\partial\beta_i}\right|_{\psi=\psi^*} = -H(q^{-1})^{-1}\left.\frac{\partial H(q^{-1},\psi)}{\partial\beta_i}\right|_{\psi=\psi^*} e(t) \tag{6.33}$$

where e_i denotes the i:th unit vector. (Note that $\varepsilon(t,\psi^*) = e(t)$!). Clearly the gradient w.r.t. θ will consist of two independent parts, one being a filtered input and the other filtered noise. The gradient w.r.t. β is independent of the input. To facilitate the forthcoming calculations introduce the notations

$$\varepsilon_\theta(t) = \left.\frac{\partial\varepsilon(t,\psi)}{\partial\theta}\right|_{\psi=\psi^*} = \varepsilon_\theta^u(t)+\varepsilon_\theta^e(t) \tag{6.34}$$

$$\varepsilon_\beta(t) = \left.\frac{\partial\varepsilon(t,\psi)}{\partial\beta}\right|_{\psi=\psi^*}$$

In (6.34) $\varepsilon_\theta^u(t)$ denotes the part depending on the input. It is readily found from (6.32) to be

$$\varepsilon_\theta^u(t) = -H(q^{-1})^{-1}\tilde{\phi}^T(t) \tag{6.35}$$

Introduce further the notations

$$\begin{aligned}
Q_{\theta\theta} &= E\varepsilon_\theta^T(t)\Lambda^{-1}\varepsilon_\theta(t)\\
Q_{\theta\theta}^u &= E\varepsilon_\theta^{uT}(t)\Lambda^{-1}\varepsilon_\theta^u(t)\\
Q_{\theta\theta}^e &= E\varepsilon_\theta^{eT}(t)\Lambda^{-1}\varepsilon_\theta^e(t)\\
Q_{\theta\beta} &= E\varepsilon_\theta^T(t)\Lambda^{-1}\varepsilon_\beta(t)\\
Q_{\beta\theta} &= Q_{\theta\beta}^T
\end{aligned} \tag{6.36}$$

$$Q_{\beta\beta} = E\varepsilon_{\beta}^{T}(t)\Lambda^{-1}\varepsilon_{\beta}(t)$$

$$\bar{Q} = Q_{\theta\theta}^{e} - Q_{\theta\beta}Q_{\beta\beta}^{-1}Q_{\beta\theta} \tag{6.37}$$

We then have, cf (6.15) and (6.35)

$$Q_{\theta\theta}^{u} = [P_{IV}^{opt}]^{-1} \tag{6.38}$$

Further since the matrix

$$\begin{bmatrix} Q_{\theta\theta}^{e} & Q_{\theta\beta} \\ Q_{\beta\theta} & Q_{\beta\beta} \end{bmatrix} = E \begin{bmatrix} \varepsilon_{\theta}^{eT}(t) \\ \varepsilon_{\beta}^{T}(t) \end{bmatrix} \Lambda^{-1}[\varepsilon_{\theta}^{e}(t) \;\; \varepsilon_{\beta}(t)]$$

is obviously nonnegative definite we have

$$\bar{Q} \geq 0 \tag{6.39}$$

Consider now in particular the model structure M_8. Then it can easily be seen directly from (6.27) that we in such a case must have

$$\varepsilon_{\theta}^{e}(t) = 0 \tag{6.40}$$

and thus

$$Q_{\theta\theta}^{e} = 0 \qquad Q_{\theta\beta} = 0 \qquad \bar{Q} = 0 \tag{6.41}$$

Return now to the covariance matrix P_{PEM}, (6.30). Using the above notations it can be written as

$$P_{PEM} = \begin{bmatrix} (P_{IV}^{opt})^{-1} + Q_{\theta\theta}^{e} & Q_{\theta\beta} \\ Q_{\beta\theta} & Q_{\beta\beta} \end{bmatrix}^{-1} \tag{6.42a}$$

In particular we get for the $\hat{\theta}$ parameters

$$\mathrm{Cov}(\hat{\theta}_{PEM}) = [(P_{IV}^{opt})^{-1} + \bar{Q}]^{-1} \leq P_{IV}^{opt} \tag{6.42b}$$

This inequality was quite expected. Note, however, that for the model structure M_8, (6.27), we find using (6.41) that the matrix P_{PEM} is block diagonal and that *equality* applies in (6.42 b). For this model parameterization, where ***independent*** parameters are used to describe the system dynamics and the noise correlation, we thus find that the optimal IV method gives the same covariance matrix as the prediction error method. Note that the results derived above on the accuracy of PEM will also be useful a little later when we will assess the asymptotic efficiency of a practical estimation procedure based on the optimal IV estimate.

Approximate implementation of the optimal IVM

We will finish this section by discussing how to approximate the optimal IV estimate given by (6.19). The other optimal IV estimate corresponding to (6.20) can be treated similarly. As already said approximations are needed for practical implementations since $\tilde{\phi}(t)$, $H(q^{-1})$ and Λ are not known in advance. In what follows we will describe a multistep algorithm for the approximate implementation. The asymptotic properties of this method will be analysed in the next section. In particular we will show that it is possible in 3 steps only to obtain an estimate $\hat{\theta}$ that is not only consistent but also asymptotically gaussian distributed with a covariance matrix equal to P_{IV}^{opt}.

The approximate multistep implementation of the optimal IV method can be described as follows. The general model structure M_6, (6.21) is used.

Approximate implementation of the optimal IV estimator

Step 1 Apply an "arbitrary" IVM in the model structure

$$y(t) = \phi^T(t)\theta+\varepsilon(t)$$

The resulting estimate will be denoted $\hat{\theta}_1$.

Step 2 Apply a PEM to estimate β (and Λ in the multivariable case) in the model structure

$$y(t) = \phi^T(t)\hat{\theta}_1+H(q^{-1},\hat{\theta}_1,\beta)\varepsilon(t,\hat{\theta}_1,\beta)$$

(with fixed $\hat{\theta}_1$). The resulting estimate will be denoted $\hat{\beta}_2$, $\hat{\Lambda}_2$.

Step 3 Compute the optimal IV estimate as given by (6.19) using $\hat{\theta}_1$ to form $\tilde{\phi}(t)$, $\hat{\theta}_1$, $\hat{\beta}_2$ to form $H(q^{-1})$, and substituting Λ by $\hat{\Lambda}_2$. The resulting estimate will be denoted $\hat{\theta}_3$.

Step 4	If desired, step 2 can be repeated using $\hat{\theta}_3$ instead of $\hat{\theta}_1$. The resulting estimate will be denoted $\hat{\beta}_4$, $\hat{\Lambda}_4$.

We may remark that within the four step algorithm we can use any method, not only a PEM, to estimate β and Λ. For instance we may use a pseudo-linear regression (PLR) algorithm which is numerically simpler than a PEM (see Stoica et al (1983) for a description of the (off-line) PLR procedure). However the accuracy of the final estimates obtained with the multistep algorithm is then expected to deteriorate. The analysis in Stoica and Söderström (1983) has shown that this is precisely what happens if we use a PLR algorithm in Steps 2 and 4 instead of the PEM.

We may also remark that if the noise does not need to be modelled we then can take $H(q^{-1}) \equiv I$ and skip Steps 2 and 4. If we repeat Step 3 using the last available estimate to form $\tilde{\phi}(t)$ we will arrive precisely at the boostrap algorithm BE_1, see (3.40)-(3.43a). The BE_1 algorithm can thus be seen as a "simplified" version of the above implementation of the optimal IV estimator.

An algorithm closely related to the multistep procedure was proposed by Young (1976), Young and Jakeman (1979, 1980), see also Jakeman and Young (1981). That algorithm consists of a repeated use of Steps 2 and 3. For the structure (6.28) of the model it was derived starting from the likelihood equations. For the structure (6.25), for which the maximum likelihood approach fails to produce algorithms of the IV-type, it was motivated using heuristics and a good engineering intuition, Jakeman and Young (1981).

6.4 ANALYSIS OF THE APPROXIMATE IMPLEMENTATION OF OPTIMAL IV ESTIMATORS

In this section we will study the accuracy properties of the approximate implementation of the optimal IV estimate which we introduced in the last section. It should be clear that all the estimates, viz. $\hat{\theta}_1$, $\hat{\beta}_2$, $\hat{\theta}_3$, $\hat{\beta}_4$ are consistent by construction. The analysis will be carried out for the general model structure M_6,(6.21). We will make frequent use of the notations $Q_{\theta\theta}$ etc, see (6.36).

The asymptotic distribution of $\hat{\theta}_1$ is of course given by Theorem 5.1. To find the accuracy of $\hat{\beta}_2$ we must take into account that $\hat{\theta}_1$ deviates from the true value θ^*. Since we are interested in asymptotic properties only, and since $\hat{\theta}_1$ is consistent we can assume that the deviation $\hat{\theta}_1-\theta^*$ is small. By continuity it then follows that $\hat{\beta}_2$ must be close to β^* since for $\hat{\theta}_1 = \theta^*$, and $N \to \infty$ we will get $\hat{\beta}_2 = \beta^*$.

As $\hat{\beta}$ denotes the minimum of $V^N(\hat{\theta}_1,\beta)$, see (6.29), we find that $\hat{\beta}_2$ satisfies

$$0 = [\frac{\partial V^N}{\partial \beta}(\hat{\theta}_1,\beta)|_{\beta=\hat{\beta}_2}]^T \approx [\frac{\partial V^N}{\partial \beta}(\theta^*,\beta)|_{\beta=\beta^*}]^T + \frac{\partial^2 V^\infty}{\partial \beta^2}(\theta^*,\beta)|_{\beta=\beta^*}(\hat{\beta}_2-\beta^*)$$

$$+ \frac{\partial^2 V^\infty}{\partial \beta \partial \theta}(\theta,\beta)|_{\theta=\theta^*,\beta=\beta^*}(\hat{\theta}_1-\theta^*) \tag{6.43}$$

where we have dropped all higher order terms.

To proceed we need the derivatives of the loss function. Some calculations give, cf Caines and Ljung (1976), (note that the evaluation is always for ψ^* and that $\varepsilon(t,\theta^*,\beta^*) = e(t)$ which is independent of all past data)

$$\frac{\partial V^N}{\partial \beta}(\theta^*,\beta)|_{\beta=\beta^*} \approx [\det(\Lambda)]\frac{2}{N}\sum_{t=1}^{N} \varepsilon^T(t,\psi^*)\Lambda^{-1}\frac{\partial \varepsilon(t,\theta^*,\beta)}{\partial \beta}|_{\beta=\beta^*} \tag{6.44a}$$

$$\frac{\partial^2 V^\infty}{\partial \beta^2}(\theta^*,\beta)|_{\beta=\beta^*} = 2[\det(\Lambda)]Q_{\beta\beta} \tag{6.44b}$$

$$\frac{\partial^2 V^\infty}{\partial \beta \partial \theta}(\theta,\beta)|_{\psi=\psi^*} = 2[\det(\Lambda)]Q_{\beta\theta} \tag{6.44c}$$

We thus get, cf (6.43), (6.44)

$$\sqrt{N}(\hat{\beta}_2-\beta^*) \approx -Q_{\beta\beta}^{-1}\{Q_{\beta\theta}\sqrt{N}(\hat{\theta}_1-\theta^*)+\frac{1}{\sqrt{N}}\sum_{t=1}^{N}[\frac{\partial \varepsilon(t,\theta^*,\beta)}{\partial \beta}|_{\beta=\beta^*}]^T\Lambda^{-1}e(t)\} \tag{6.45}$$

Note that the same relation holds if we make the substitutions

$$\hat{\theta}_1 \to \hat{\theta}_3,\ \hat{\beta}_2 \to \hat{\beta}_4.$$

In view of Assumption A8 ($Z(\cdot)$ and $v(\cdot)$ independent) the two terms of (6.45) are asymptotically uncorrelated. This can be shown as follows (assuming for simplicity that $nz = n\theta$ in Step 1)

$$E\sqrt{N}(\hat{\theta}_1-\theta^*)\frac{1}{\sqrt{N}}\sum_{t=1}^{N} e^T(t)\Lambda^{-1}\frac{\partial \varepsilon(t,\theta^*,\beta)}{\partial \beta}|_{\beta=\beta^*} \tag{6.46}$$

$$\approx [EZ(t)\cdot F(q^{-1})\phi^T(t)]^{-1}E\frac{1}{N}\sum_{s,t=1}^{N} Z(s)\cdot F(q^{-1})v(s)\cdot e^T(t)\Lambda^{-1}\frac{\partial \varepsilon(t,\theta^*,\beta)}{\partial \beta}|_{\beta=\beta^*} = 0$$

The reason for the last equality is that Z(t) depends on the input while the other factors are obtained by filtering the noise. It is now possible to establish the asymptotic distribution of the estimates determined in Step 1 and Step 2.

Theorem 6.2 Suppose that the Assumptions

A2, A4, A5 (existence of an asymptotically stable system corresponding to a unique θ^*)

A3, A6, A7 ($u(\cdot)$, $v(\cdot)$ stationary and independent)

A8 ($Z(\cdot)$ independent of $v(\cdot)$)

all apply. Consider the estimates $\hat{\theta}_1$ and $\hat{\beta}_2$ of the multistep algorithm described in Section 6.3. The estimates are asymptotically gaussian distributed

$$\sqrt{N}\begin{bmatrix}\hat{\theta}_1-\theta^* \\ \hat{\beta}_2-\beta^*\end{bmatrix} \xrightarrow{dist} N(0,P_2) \tag{6.47a}$$

where

$$P_2 = \begin{bmatrix} P_\theta^{(1)} & P_{\theta\beta}^{(12)} \\ P_{\beta\theta}^{(12)} & P_\beta^{(2)} \end{bmatrix} \tag{6.47b}$$

$$P_\theta^{(1)} = [EZ(t)\cdot F(q^{-1})\phi^T(t)]^{-1}E\{[\sum_{i=0}^{\infty} Z(t+i)K_i]\Lambda[\sum_{j=0}^{\infty} K_j^T Z^T(t+j)]\}$$

$$\cdot[EF(q^{-1})\phi(t)\cdot Z^T(t)]^{-1} \tag{6.47c}$$

$$P_{\theta\beta}^{(12)} = -P_\theta^{(1)}Q_{\theta\beta}Q_{\beta\beta}^{-1} = P_{\beta\theta}^{(12)T} \tag{6.47d}$$

$$P_\beta^{(2)} = Q_{\beta\beta}^{-1}[Q_{\beta\theta}P_\theta^{(1)}Q_{\theta\beta}+Q_{\beta\beta}]Q_{\beta\beta}^{-1} \tag{6.47e}$$

Proof. The basic result (6.47a) follows in the manner of Theorem 5.1. The expression (6.47c) follows directly from Theorem 5.1. Further, (6.47d,e) follow from (6.45), (6.46) after noticing that

$$\lim_{N\to\infty} \frac{1}{N}\sum_{t,s=1}^{N} E\frac{\partial\varepsilon}{\partial\beta}^{T}(t)\Lambda^{-1}e(t)e^{T}(s)\Lambda^{-1}\frac{\partial\varepsilon}{\partial\beta}(s) = Q_{\beta\beta}$$ ∎

Remark: Note that the first term in (6.47e) shows how the uncertainities in $\hat{\beta}_2$ increase due to the deviation of $\hat{\theta}_1$ from θ^*. If we put $P_{\theta}^{(1)} = 0$ in (6.47e) we would get $P_{\beta}^{(2)} = Q_{\beta\beta}^{-1}$, which is in accordance with (6.36), (6.42a). ∎

We then proceed to examine the accuracy of the estimate $\hat{\theta}_3$ as given by (6.19) where the matrix Λ, the filter $H(q^{-1})$ and the matrix $\tilde{\phi}(t)$ are replaced with some consistent estimates:

$$\hat{\theta}_3 = \{\sum_{t=1}^{N} [\hat{H}(q^{-1})^{-1}\hat{\tilde{\phi}}^{T}(t)]^{T}\hat{\Lambda}^{-1}[\hat{H}(q^{-1})^{-1}\phi^{T}(t)]\}^{-1}$$

$$\cdot\{\sum_{t=1}^{N} [\hat{H}(q^{-1})^{-1}\hat{\tilde{\phi}}^{T}(t)]^{T}\hat{\Lambda}^{-1}[\hat{H}(q^{-1})^{-1}y(t)]\} \tag{6.48}$$

It is trivial to see that since $\hat{\tilde{\phi}}(t)$ is a consistent estimate of $\tilde{\phi}(t)$, the inverse appearing in (6.48) will exist at least for sufficiently large N. Furthermore, some straightforward calculations show that for (6.48) we have asymptotically

$$\hat{\theta}_3-\theta^* = \{\sum_{t=1}^{N} [\hat{H}(q^{-1})^{-1}\hat{\tilde{\phi}}^{T}(t)]^{T}\hat{\Lambda}^{-1}[\hat{H}(q^{-1})^{-1}\phi^{T}(t)]\}^{-1}$$

$$\cdot\{\sum_{t=1}^{N} [\hat{H}(q^{-1})^{-1}\hat{\tilde{\phi}}^{T}(t)]^{T}\hat{\Lambda}^{-1}[\hat{H}(q^{-1})^{-1}H(q^{-1})e(t)]\}$$

$$= \{E[H(q^{-1})^{-1}\tilde{\phi}^{T}(t)]^{T}\Lambda^{-1}[H(q^{-1})^{-1}\tilde{\phi}^{T}(t)]\}^{-1}$$

$$\cdot\{\frac{1}{N}\sum_{t=1}^{N} [H(q^{-1})^{-1}\tilde{\phi}^{T}(t)]^{T}\Lambda^{-1}e(t)\}\{1+o(1)\} \tag{6.49}$$

The use of $\hat{H}(q^{-1})$ and $\hat{\tilde{\phi}}(t)$ instead of $H(q^{-1})$ and $\tilde{\phi}(t)$ gives hence only *higher* order effects, so that the IV estimate (6.48) is still asymptotically optimal. This result is a key point in analysing the accuracy of the estimates obtained in Steps 2 & 3 and 3 & 4. The asymptotic distributions for these estimates are given in the following theorem.

Theorem 6.3. Suppose that the Assumptions

A2, A4, A5 (existence of an asymptotically stable system corresponding to a unique θ^*)

A3, A6, A7 ($u(\cdot)$, $v(\cdot)$ stationary and independent)

A8 ($Z(\cdot)$ independent of $v(\cdot)$)

all apply. Consider the estimates $\hat{\beta}_2$, $\hat{\theta}_3$ and $\hat{\beta}_4$ of the multistep algorithm described in Section 6.3. They are asymptotically gaussian distributed

$$\sqrt{N}\begin{bmatrix}\hat{\theta}_3-\theta^* \\ \hat{\beta}_2-\beta^*\end{bmatrix} \xrightarrow{\text{dist}} N(0,P_3) \tag{6.50a}$$

$$P_3 = \begin{bmatrix} P_\theta^{(3)} & P_{\theta\beta}^{(23)} \\ P_{\beta\theta}^{(23)} & P_\beta^{(2)} \end{bmatrix} \tag{6.50b}$$

$$\sqrt{N}\begin{bmatrix}\hat{\theta}_3-\theta^* \\ \hat{\beta}_4-\beta^*\end{bmatrix} \xrightarrow{\text{dist}} N(0,P_4) \tag{6.51a}$$

$$P_4 = \begin{bmatrix} P_\theta^{(3)} & P_{\theta\beta}^{(34)} \\ P_{\beta\theta}^{(34)} & P_\beta^{(4)} \end{bmatrix} \tag{6.51b}$$

The matrices involved are given by

$$P_\theta^{(3)} = P_{IV}^{opt} \quad \text{(see (6.15))} \tag{6.52a}$$

$$P_{\theta\beta}^{(23)} = P_{\theta\beta}^{(34)} = -P_\theta^{(3)}Q_{\theta\beta}Q_{\beta\beta}^{-1} = P_{\beta\theta}^{(23)T} = P_{\beta\theta}^{(34)T} \tag{6.52b}$$

$$P_\beta^{(4)} = Q_{\beta\beta}^{-1}[Q_{\beta\theta}P_\theta^{(3)}Q_{\theta\beta}+Q_{\beta\beta}]Q_{\beta\beta}^{-1} \tag{6.52c}$$

Further, $P_{\beta}^{(2)}$ is given by (6.47e).

Proof. The convergence to gaussian distribution follows as in Theorem 5.1. It remains to verify the covariance matrices (6.52).

The expression (6.52a) follows from (6.49). To derive (6.52b), notice that from (6.45), (6.49) we get

$$P_{\theta\beta}^{(23)} = -P_{IV}^{opt} \lim_{N\to\infty} E \frac{1}{\sqrt{N}} \sum_{t=1}^{N} \{H(q^{-1})^{-1}\tilde{\phi}^{T}(t)\}^{T}\Lambda^{-1}e(t)\cdot\sqrt{N}\ (\hat{\theta}_{1}-\theta^{*})^{T}Q_{\theta\beta}Q_{\beta\beta}^{-1}$$

$$= -P_{\theta}^{(3)}[\lim_{N\to\infty}\frac{1}{N}\sum_{s,t=1}^{N}\sum_{i=0}^{\infty} E\{H(q^{-1})^{-1}\tilde{\phi}^{T}(t)\}^{T}\Lambda^{-1}e(t)\{K_{i}e(s-i)\}^{T}Z^{T}(s)]$$

$$\cdot[E\{F(q^{-1})\phi^{T}(t)\}^{T}Z^{T}(t)]^{-1}Q_{\theta\beta}Q_{\beta\beta}^{-1}$$

$$= -P_{\theta}^{(3)}[E\{H(q^{-1})^{-1}\tilde{\phi}^{T}(t)\}^{T}\{\sum_{i=0}^{\infty} K_{i}^{T}Z^{T}(t+i)\}]$$

$$\cdot[E\{F(q^{-1})\tilde{\phi}^{T}(t)\}^{T}Z^{T}(t)]^{-1}Q_{\theta\beta}Q_{\beta\beta}^{-1} \qquad (6.53)$$

However,

$$E[\{H(q^{-1})^{-1}\tilde{\phi}^{T}(t)\}^{T}\{\sum_{i=0}^{\infty} K_{i}^{T}Z^{T}(t+i)\}] = E\{\sum_{i=0}^{\infty} K_{i}q^{-i}H(q^{-1})^{-1}\tilde{\phi}^{T}(t)\}^{T}Z^{T}(t)$$

$$= E[F(q^{-1})\tilde{\phi}^{T}(t)]^{T}Z^{T}(t)$$

and thus the expression (6.52b) of $P_{\theta\beta}^{(23)}$ results easily.

Finally the first equality in (6.52b) as well as (6.52c) follow from the remark made after (6.45) and (6.47d) respectively (6.47e). ∎

We have now derived the asymptotic distributions of the parameter estimates obtained within the multistep algorithm. Interestingly enough, it can be seen that the asymptotic distribution of the estimates will remain unchanged if further steps are added to the algorithm. We can thus expect that for a sufficiently large number of data *the multistep algorithm will converge in four steps only*. The results on the asymptotic distribution can also be used to compare the accuracy reached after the different steps. The covariance matrices of interest in this respect are P_2, (6.47b), P_3, (6.50b), and P_4, (6.51b).

Note that from Theorem 6.1 we have

$$P_\theta^{(1)} \geq P_\theta^{(3)} = P_{IV}^{opt} \tag{6.54}$$

and thus also

$$P_\beta^{(2)} \geq P_\beta^{(4)} \tag{6.55}$$

Hence we trivially have tr $P_2 \geq$ tr P_3, for instance. However, the stronger order relation $P_2 \geq P_3$ will in general not hold since in P_2-P_3 the 22-block is zero while the 12-block in most cases is nonzero. Interestingly enough, such a strong relation can be shown to exist between P_2 and P_4, and P_3 and P_4, respectively.

Lemma 6.1. Consider the covariance matrices P_2, (6.47b), P_3, (6.50b), and P_4, (6.51b). Then

$$P_2 \geq P_4 \tag{6.56}$$

$$P_3 \geq P_4 \tag{6.57}$$

Proof. We get easily from (6.47), (6.51)

$$P_2-P_4 = \{\begin{bmatrix} I \\ -Q_{\beta\beta}^{-1}Q_{\beta\theta} \end{bmatrix} P_\theta^{(1)} [I \;\; -Q_{\theta\beta}Q_{\beta\beta}^{-1}] + \begin{bmatrix} 0 & 0 \\ 0 & Q_{\beta\beta}^{-1} \end{bmatrix}\}$$

$$-\{\begin{bmatrix} I \\ -Q_{\beta\beta}^{-1}Q_{\beta\theta} \end{bmatrix} P_\theta^{(3)} [I \;\; -Q_{\theta\beta}Q_{\beta\beta}^{-1}] + \begin{bmatrix} 0 & 0 \\ 0 & Q_{\beta\beta}^{-1} \end{bmatrix}\}$$

$$= \begin{bmatrix} I \\ -Q_{\beta\beta}^{-1}Q_{\beta\theta} \end{bmatrix} (P_\theta^{(1)} - P_\theta^{(3)}) [I \;\; -Q_{\theta\beta}Q_{\beta\beta}^{-1}]$$

which shows (6.56), see (6.54). Next using (6.50)-(6.53) we get

$$P_3-P_4 = \begin{bmatrix} 0 & 0 \\ 0 & P_\beta^{(2)}-P_\beta^{(4)} \end{bmatrix}$$

and thus (6.57) follows from (6.55). ∎

The result of Lemma 6.1 is simply that it pays off (in terms of accuracy) to perform the last two steps of the algorithm.

We next give a more detailed result on the relation between P_2, P_3 and P_4 that holds whenever $Q_{\theta\beta} = 0$.

Lemma 6.2. Consider the matrices P_2, (6.47b), P_3, (6.50b), and P_4, (6.51b). Assume that the model structure is such that $Q_{\theta\beta} = 0$. Then

i) P_2, P_3 and P_4 are all block diagonal

ii) $P_2 \geq P_3 = P_4$ (6.58)

Proof. Part i) is immediate from (6.47d), (6.52b). What remains to show in order to prove (6.58) is only that $P_\beta^{(2)} = P_\beta^{(4)}$. However, this follows directly from (6.47e), (6.52c). ∎

The result of Lemma 6.2 is strong and neat. Note that the assumption $Q_{\theta\beta} = 0$ is fulfilled for the model structure M_8,(6.27) (as well as for its special case M_8', (6.28)). This was shown in (6.40), (6.41). For such a model structure we get maximal accuracy after 3 steps only. Moreover, the estimates $\hat{\theta}$ and $\hat{\beta}$ are (asymptotically) uncorrelated.

It is also of great interest to compare the accuracy of the multistep method with that of a prediction error method. This is done in the next lemma.

Lemma 6.3. Consider the matrices P_4, (6.51b) and P_{PEM}, (6.30). Then

$$P_4 \geq P_{PEM} \tag{6.59}$$

Equality in (6.59) holds if and only if

$$\bar{Q} = 0 \tag{6.60}$$

($\bar{Q}$ was defined in (6.37)).

<u>*Proof.*</u> Consider the following calculations

$$P_{PEM}^{-1}(P_4-P_{PEM})P_{PEM}^{-1} = \begin{bmatrix} Q_{\theta\theta} & Q_{\theta\beta} \\ Q_{\beta\theta} & Q_{\beta\beta} \end{bmatrix} \begin{bmatrix} I \\ -Q_{\beta\beta}^{-1}Q_{\beta\theta} \end{bmatrix} P_{IV}^{opt}[I \; -Q_{\theta\beta}Q_{\beta\beta}^{-1}] \begin{bmatrix} Q_{\theta\theta} & Q_{\theta\beta} \\ Q_{\beta\theta} & Q_{\beta\beta} \end{bmatrix}$$

$$+\begin{bmatrix} Q_{\theta\beta} \\ Q_{\beta\beta} \end{bmatrix} Q_{\beta\beta}^{-1}[Q_{\beta\theta} \quad Q_{\beta\beta}] - \begin{bmatrix} Q_{\theta\theta} & Q_{\theta\beta} \\ Q_{\beta\theta} & Q_{\beta\beta} \end{bmatrix} = \begin{bmatrix} I \\ 0 \end{bmatrix} (\bar{Q}+\bar{Q}P_{IV}^{opt}\bar{Q})[I \quad 0].$$

Since $\bar{Q}$ and P_{IV}^{opt} are nonnegative definite (6.59) follows. Moreover, equality in (6.59) is equivalent to

$$\bar{Q}+\bar{Q}P_{IV}^{opt}\bar{Q} = 0$$

which trivially can be simplified to $\bar{Q} = 0$. The lemma is thus proved. ■

The important conclusion is that <u>*for the model structure M_8 with independent parameters describing the system dynamics and the noise correlation the multistep algorithm will not only give convergence in three steps (at least asymptotically) but it will also give the same accuracy as a prediction error method, cf. (6.41) and Lemma 6.3. However, for other model structures, the condition (6.60) is in general not fulfilled*</u>. We will give an example of this.

<u>*Example 6.7*</u> *Examination of the condition $\bar{Q} = 0$*

Consider the SISO model structure (6.25), i.e.

$$A(q^{-1})y(t) = B(q^{-1})u(t)+ \frac{C(q^{-1})}{D(q^{-1})} \varepsilon(t,\theta,\beta) \tag{6.61}$$

Assume that the true system is given by

$$A^*(q^{-1})y(t) = B^*(q^{-1})u(t)+\frac{C^*(q^{-1})}{D^*(q^{-1})}e(t) \tag{6.62}$$

Some straightforward calculation then gives, using the notations of (6.34)

$$\varepsilon_\theta^e(t) = \frac{1}{A^*(q^{-1})}[e(t-1)\ldots e(t-na) \mid \underbrace{0\ldots 0}_{nb\ \text{elements}}]$$

$$\varepsilon_\beta(t) = [-\frac{1}{C^*(q^{-1})}e(t-1)\ldots -\frac{1}{C^*(q^{-1})}e(t-nc)\ \frac{1}{D^*(q^{-1})}e(t-1)\ldots \frac{1}{D^*(q^{-1})}e(t-nd)]$$

It is clear that in general $Q_{\theta\beta} = E\varepsilon_\theta^T(t)\varepsilon_\beta(t)$ is nonzero. Thus four steps are needed to get the maximal accuracy of the multistep algorithm, cf e.g. Lemma 6.1. Consider then the condition $\bar{Q} = 0$. It follows from Lemma A3.9 that $\bar{Q} = 0$ is equivalent to

$$\text{rank}\, E\begin{bmatrix}\varepsilon_\theta^{e\,T}(t)\\ \varepsilon_\beta^T(t)\end{bmatrix}[\varepsilon_\theta^e(t)\quad \varepsilon_\beta(t)] = nc+nd \tag{6.63}$$

However, the matrix appearing in (6.63) is generically of rank na+nc+nd. This can be seen as follows. Using the approach taken in the proof of Lemma A3.1 we find that the matrix has lower rank if and only if there exist polynomials

$$\tilde{A}(z) = \sum_{i=1}^{na}\tilde{a}_i z^i \qquad \tilde{C}(z) = \sum_{i=1}^{nc}\tilde{c}_i z^i \qquad \tilde{D}(z) = \sum_{i=1}^{nd}\tilde{d}_i z^i,$$

at least one being nonzero, such that

$$\frac{\tilde{A}(z)}{A^*(z)} \equiv \frac{\tilde{D}(z)C^*(z)-D^*(z)\tilde{C}(z)}{D^*(z)C^*(z)} \tag{6.64}$$

Assume that the polynomials $A^*(z)$, $C^*(z)$ and $D^*(z)$ are pairwise coprime. This is generically true. Then the two sides of (6.64) cannot have any common pole, so all poles must be cancelled by zeros. However, the degrees of the numerators are less than the degrees of the corresponding denominators. Hence the above identity implies easily $\tilde{A}(z) \equiv \tilde{C}(z) \equiv \tilde{D}(z) \equiv 0$. Hence a contradiction to $\bar{Q} = 0$ is obtained. ∎

We have now seen that the model structure used is crucial when comparing the accuracy that can be achieved with the multistep method with that of a prediction error method.

When the system dynamics and the noise correlation are described with different parameters, then these two methods give the same accuracy (see Lemma 6.3 and the following discussion). However, for model parameterizations not fulfilling the above condition a prediction error method will in general give superior accuracy.

A further evaluation of the multistep algorithm is presented in the following numerical example.

Example 6.8 Numerical evaluation of the multistep algorithm

The aim of this example is to discuss the following problems.

- Assume that one aims at estimating only the parameters of the system dynamics (i.e. the transfer function parameters). There is then a good reason to skip Steps 2 and 4 of the multistep algorithm and to take as Step 2 the IV estimate (6.48) with $\hat{H}(q^{-1}) \equiv I$. As mentioned before the BE_1 algorithm will then be obtained. It is computationally simpler but certainly less accurate than the multistep algorithm. Some comparisons in more quantitative terms would, however, be valuable.

- For the model M_7^1, (6.25), the multistep algorithm is inferior from the accuracy point of view to a PEM. The difference should, however, be investigated in more detail.

- The multistep algorithm achieves, asymptotically, maximum accuracy in at most four steps. In finite samples it might be beneficial to re-iterate Steps 3 and 4, cf Young and Jakeman (1979 a). That such a repeated use of Steps 3 and 4 will lead, in finite samples, to an increase of accuracy is, however, only a conjecture which should be further investigated.

The following system

$$A^*(q^{-1})y(t) = B^*(q^{-1})u(t) + \frac{1}{D(q^{-1})} e(t) \tag{6.65}$$

with

$$A^*(q^{-1}) = 1-0.3q^{-1}+0.5q^{-2}$$

$$B^*(q^{-1}) = 1.5q^{-1}$$

$$D(q^{-1}) = 1-0.7q^{-1}+0.9q^{-2}$$

was simulated. Twenty different realizations, each of length N = 600 were generated. The input u(t) was a sequence of independent and identically distributed random variables with zero mean and unit variance. The values S = 1 and S = 5 of the signal-to-noise ratio were considered in the simulations.

Assuming that the orders of the system are known, the parameters of (6.65) were estimated using the following procedures.

- The multi (four)-step algorithm [4SA]

- A ten-step algorithm [10SA] obtained by appropriately reiterating the last two steps of 4SA.

- The two-step algorithm [(4/2)SA] = BE_1

- A five-step algorithm [(10/2)SA] obtained by reiterating the last step of (4/2)SA.

The first step for all algorithms is a simple IV method with

$$z(t) = [u(t-1)\dots\ u(t-na-nb)]^T$$

The first two algorithms estimate also a noise model (in Steps 2, 4 etc.). Note, however, that for the considered structure of system this is an ordinary least-squares problem.

The results presented in Table 6.1 are means and standard deviations obtained from 20 realizations. The table includes also sample estimates of the (asymptotic) standard deviations corresponding to a PEM. The results would suggest the following.

- The 4SA gives considerably better accuracy than (4/2)SA. The difference between the CPU times required by these two methods is small and should not be too important for off-line applications.

- The estimate obtained with the 4SA and that corresponding to the 10SA are very close to each other. The same is true for (4/2)SA and (10/2)SA. This was expected in view of the theory previously developed. It should, however, be stressed once more here that the theory concerns asymptotic properties. For the present example with N = 600 these properties apply. For other systems, however, even with longer samples it might be necessary to reiterate the last step(s) of the 4SA [or (4/2)SA]

Parameters	True values	Estimated values									
		S = 5					S = 1				
		(4/2)SA $=BE_1$	(10/2)SA	4SA	10SA	PEM (st.dev.)	(4/2)SA $=BE_1$	(10/2)SA	4SA	10SA	PEM (st.dev.)
a_1	-0.3	-0.301± 0.006	-0.301± 0.002	-0.300± 0.002	-0.300± 0.002	0.002	-0.306± 0.029	-0.306± 0.029	-0.301± 0.012	-0.301± 0.012	0.012
a_2	0.5	0.499± 0.007	0.499± 0.007	0.500± 0.002	0.500± 0.002	0.002	0.497± 0.035	0.498± 0.035	0.498± 0.012	0.498± 0.012	0.011
b_1	1.5	1.502± 0.008	1.502± 0.008	1.501± 0.004	1.501± 0.003	0.003	1.512± 0.037	1.512± 0.037	1.503± 0.019	1.503± 0.018	0.016
d_1	-0.7	-	-	-0.694± 0.016	-0.694± 0.016	0.018	-	-	-0.694± 0.016	-0.694± 0.016	0.019
d_2	0.9	-	-	0.891± 0.018	0.891± 0.018	0.018	-	-	0.891± 0.019	0.891± 0.019	0.019

Table 6.1. Simulation results based on 20 realizations each of length N = 600. The figures shown are arithmetic means and standard deviations.

in order to improve the estimates. In most cases it is though, according to our experience, sufficient to use 4SA [or (4/2)SA].

- For the (reasonable) values of N and S considered in this example, the 4SA and the PEM give practically the same accuracy. ■

6.5 REMARKS AND BIBLIOGRAPHICAL NOTES

A further discussion on accuracy functions, such as those introduced in Example 6.1 can be found in Gustavsson et al (1977, 1981), Ljung and Söderström (1983).

Minimax optimization for determining the design variables is further discussed in Stoica and Söderström (1981 b, 1982 b).

Theorem 6.1 is taken from Stoica and Söderström (1981 d, 1983). It can be seen both as a considerable extension and a refinement of the result on optimal IV estimation given by Wong and Polak (1967).

The approximate algorithm for computing the optimal IV estimate is essentially an off-line procedure. Both the IV method used in Steps 1, 3 etc and the PE method used in Steps 2, 4 etc can be converted to on-line algorithms. However, if we do so and also allow the transfer of estimates between steps each time a new data pair is processed (in order to get a true on-line procedure) then we shall arrive at an iterative rather than a four-step algorithm. In this book we have concentrated on the multistep algorithm. For discussions on the on-line implementation we refer the reader to Solo (1980), Young (1976), Young and Jakeman (1979 a, 1980), Jakeman and Young (1979, 1981), see also Ljung and Söderström (1983).

Algorithms more or less similar to the multistep procedure have been proposed in several papers in the econometric literature like Engle (1980), Hausman (1975), Hendry (1976), Lyttkens (1974). The model dealt with in these references is the "simultaneous equation model" which is somewhat different from our models. In this respect the references cited are less relevant for our treatment. Nevertheless some methods appearing in these references are very similar in spirit to the multistep algorithm.

Chapter 7

OPTIMAL INPUT DESIGN

7.1 STATEMENT OF THE PROBLEM

We have in Chapter 2 introduced the concepts of system S, model structure M, identification method J and experimental condition X in order to describe and analyse identification algorithms. Of these, S is given (fixed) but unknown while the other three can be considered to be at the user's disposal. We have in Chapter 2 presented several examples of how to choose the model structure M. We will give a complementary discussion on the model order selection in Section 8.4. In the previous chapters we have also studied various types of IV identification methods.

In this chapter we will consider the problem of choosing the experimental condition X. We will concentrate on the question how to design inputs that are optimal in a certain sense.

The problem of optimal input design appears not only for IVMs but also for other identification methods. We will present one convenient way of computing optimal inputs and discuss how it can be used in connection with the multistep algorithm we derived in Chapter 6. The approach taken can, though, be used also for PEMs and some other schemes. The treatment will be restricted to SISO systems.

Let us first give some general comments on the optimal input design problem.

An optimal input (i.e. an input maximizing the "model accuracy", see below for a formal definition) will, in general, depend on the true system. Similarly we saw in Chapter 6 that the optimal IV method requires knowledge of the system. This means that a priori information is needed to compute the optimal input. In practice one often has to do a preliminary experiment to get a first model. When computing the optimal input signal this model will be used as if it gives a perfect description of the system. Using the optimal input, a new (improved) model is then estimated. In contrast to the optimization of instruments we cannot use the *same* data set to determine the refined model, but a *new* identification experiment is needed. This requirement may rule out the possibility to use optimal inputs in some cases.

A general measure of the accuracy of an estimated model is given by the covariance matrix of the parameter estimates. It is in general not possible, however, to find inputs that optimize the whole covariance matrix. Instead, we have to optimize scalar functions of the covariance matrix. We gave examples of such scalar functions and called them accuracy functions in Example 6.1. To formulate the optimization problem we must also introduce a restriction on the input signal. Otherwise the optimal input will clearly be an infinite power signal making the covariance matrix of the estimated parameters equal to zero. We will in general denote such a constraint by

$$u(\cdot) \in \mathcal{D} \tag{7.1}$$

The explicit analysis will be carried out for constrained input variance,

$$\mathcal{D} = \mathcal{D}_u \triangleq \{u(\cdot) | Eu^2(t) \leq \sigma_u^2\} \tag{7.2a}$$

and for constrained output variance,

$$\mathcal{D} = \mathcal{D}_y \triangleq \{u(\cdot) | Ex^2(t) \leq \sigma_x^2\} \tag{7.2b}$$

Recall that x(t) in (7.2b) is the noise-free output defined by (2.9).

We now have the following

Optimal input design problem (7.3a) Find an input u(·) that minimizes the scalar accuracy function f(P(u)) subject to the constraint $u(\cdot) \in \mathcal{D}$. Here P(u) denotes the covariance matrix P_4, (6.51), of the estimates obtained with the multistep algorithm, and f(·) is a scalar-valued, monotonically increasing function.

Note that, as described in Example 6.1, the function f(P) can be chosen, for instance, as det(P) or tr(WP) (with W a positive definite weighting matrix). The inputs minimizing these two loss functions are usually called D-optimal and A-optimal inputs, respectively, see e.g. Mehra (1974).

Let us now make a simple observation on the optimization problem just formulated. The optimal input signal will clearly belong to the boundary of $\mathcal{D}$, such that equality will hold in (7.2 a or b). To see this, simply notice that f(P(u)) is inversely proportional to the input variance. Hence we are motivated to work with the following constraint sets, instead of (7.2)

$$\mathcal{D} = \mathcal{D}'_u \triangleq \{u(\cdot) | Eu^2(t) = \sigma_u^2\} \tag{7.4a}$$

for constrained input variance, and

$$\mathcal{D} = \mathcal{D}_y' \triangleq \{u(\cdot)\,|\,Ex^2(t) = \sigma_x^2\} \qquad (7.4b)$$

for constrained output variance.

In the following section we will describe the inputs by using a parameter vector ρ. The input could then be denoted $u_\rho(t)$. The idea behind the introduction of ρ is to convert the optimal input design problem into a standard nonlinear programming task. Let R be the allowable set for ρ.

As an example, let $u_\rho(t)$ be an ARMA process of order n. Then ρ will consist of the 2n polynomial coefficients describing the ARMA process. A restriction must clearly be used to guarantee stability of the process. Also, to ensure a unique representation, the zeros of the ARMA model should be constrained to lie within the unit circle as well. These constraints define the allowable set R in this case.

We can now restate the optimal input determination problem using the parameterized input.

Reformulated optimal input design problem (7.3b)
Find a parameter vector ρ that minimizes $f(P(u_\rho))$ subject to the constraint $\rho \in \bar{\mathcal{D}} \cap R$, where $\bar{\mathcal{D}} = \{\rho\,|\,u_\rho(\cdot) \in \mathcal{D}\}$

The choice of parameters for describing the input is of course a crucial point. We will discuss a specific parameterization in the next section. Let us here only point out that in order to guarantee that an over-all optimal input is obtained (i.e. no better result can be obtained with any other parameterization) it is most natural to require that for *any* $\tilde{u}(\cdot) \in \mathcal{D}$ there is a $\rho \in R$ such that

$$u_\rho(\cdot) \in \mathcal{D} \text{ and } P(u_\rho) = P(\tilde{u}) \qquad (7.5)$$

7.2 INPUT PARAMETERIZATION

We will now show that there is a simple input parameterization that fulfils (7.5). In the following sections we will discuss in some detail the algorithmic aspects of the corresponding optimization problem (7.3b).

It follows from Theorem 6.3 that the input affects the asymptotic covariance matrix of the multistep algorithm only in the block P_{IV}^{opt}. The same statement is also true for a PEM, as can be seen from (6.42a). For SISO systems we get from (6.15), neglecting a trivial factor of $\Lambda = \lambda$

$$P_{IV}^{opt} = \{E[H^{-1}(q^{-1})\tilde{\varphi}(t)][H^{-1}(q^{-1})\tilde{\varphi}^T(t)]\}^{-1} \tag{7.6a}$$

with

$$\tilde{\varphi}(t) = [-\frac{B^*(q^{-1})}{A^*(q^{-1})}u(t-1)\ldots -\frac{B^*(q^{-1})}{A^*(q^{-1})}u(t-na)\ u(t-1)\ldots u(t-nb)]^T \tag{7.6b}$$

Since $H(q^{-1})$ is a rational filter we can write it as

$$H(q^{-1}) = \frac{S(q^{-1})}{T(q^{-1})} \tag{7.7a}$$

where $S(z)$ and $T(z)$ are the following polynomials

$$S(z) = 1+s_1z+\ldots+s_{ns}z^{ns}$$
$$T(z) = 1+t_1z+\ldots+t_{nt}z^{nt} \tag{7.7b}$$

The system is thus assumed to be described by

$$A^*(q^{-1})y(t) = B^*(q^{-1})u(t)+\frac{S(q^{-1})}{T(q^{-1})}e(t) \tag{7.7c}$$

We can compare this structure with the model structures M_7' and M_8', introduced in Example 6.6. For M_7', (6.25), we get $S(z) \equiv C(z)$, $T(z) \equiv D(z)$ while M_8', (6.28), implies $S(z) \equiv A^*(z)C(z)$, $T(z) \equiv D(z)$.

In the following we will assume, for simplicity of notation, that $na = na^*$, $nb = nb^*$. Note, though, that the analysis can directly be extended to the case $\min(na-na^*, nb-nb^*) = 0$.

The matrix P_{IV}^{opt} fulfils

$$(P_{IV}^{opt})^{-1} = \sigma^2 S(-B^*,A^*)\{E\, T(q^{-1}) \begin{bmatrix} \bar{u}(t-1) \\ \vdots \\ \bar{u}(t-na-nb) \end{bmatrix}$$

$$\cdot\, T(q^{-1})[\bar{u}(t-1) \ldots \bar{u}(t-na-nb)]\} S^T(-B^*,A^*) \tag{7.8}$$

where $S(-B^*,A^*)$ is the nonsingular Sylvester matrix associated with $-B^*$ and A^*, see Lemma A3.1, and

$$\sigma^2 \triangleq E[\frac{1}{A^*(q^{-1})S(q^{-1})} u(t)]^2 \tag{7.9}$$

$$\bar{u}(t) \triangleq \frac{1}{\sigma A^*(q^{-1})S(q^{-1})} u(t) \tag{7.10}$$

It can now be easily seen that the matrix P_{IV}^{opt} (and hence $P(u)$ too) is completely determined by σ and the following $(na+nb+nt-1)$ covariance elements of $\bar{u}(t)$,

$$\rho_i \triangleq E\bar{u}(t)\bar{u}(t-i) \qquad i = 1, 2, \ldots, na+nb+nt-1 \tag{7.11}$$

Since $E\bar{u}^2(t) = 1$, the sequence $\{\rho_i\}$ can also be viewed as the autocorrelation function of $\bar{u}(t)$.

We next show that the sets $\mathcal{D}_u'$, (7.4a), and $\mathcal{D}_y'$, (7.4b), can also be described with the autocorrelations $\{\rho_i\}$. For this purpose rewrite (7.4a) and (7.4b) as

$$\mathcal{D}_u' = \{u(t) | \sigma^2 E[A^*(q^{-1})S(q^{-1})\bar{u}(t)]^2 = \sigma_u^2\} \tag{7.12}$$

and

$$\mathcal{D}_y' = \{u(t) | \sigma^2 E[B^*(q^{-1})S(q^{-1})\bar{u}(t)]^2 = \sigma_x^2\} \tag{7.13}$$

Clearly (7.12) depends on the first $na+ns$ covariances of $\bar{u}(t)$, and (7.13) on the first $nb+ns-1$ covariances.

The above discussion shows that

$$\sigma \text{ and } \{\rho_i\}_{i=1}^{p} \text{ with } p = \begin{cases} na+\max(ns,\ nb+nt-1) \text{ for } \mathcal{D}_u' \\ nb-1+\max(ns,\ na+nt) \text{ for } \mathcal{D}_y' \end{cases} \tag{7.14}$$

is a possible parameterization of the (filtered) input $\bar{u}(t)$ which allows us to formulate the optimal design problem as a standard nonlinear programming problem. Once the optimal $\bar{u}(t)$ is found, it is easy to compute $u(t)$ from (7.10). There are, however, some points to be clarified before we can use this parameterization in applications.

- How should the covariance sequence $\{\rho_i\}$ be constrained in the course of optimization?
- How should the input corresponding to the optimal covariances, say $\{\hat{\rho}_i\}$, be generated?

We shall approach such questions in the following sections.

Here we note that the optimization with respect to σ^2 is immediate. Indeed, σ^2 is obtained from the constraint (7.12) or (7.13) as a function of

$$\rho \triangleq [\rho_1 \; \rho_2 \ldots \rho_p]^T \tag{7.15}$$

Denote this function by $\hat{\sigma}^2(\rho)$. Using $\hat{\sigma}^2(\rho)$ in (7.8) and P_{IV}^{opt} so obtained in (7.3b) we arrive at a static optimization problem with $\{\rho_i\}_{i=1}^p$ as independent variables. The optimization with respect to ρ needs to be discussed in detail. This will be done in the next section. There, we will also show how the optimal input signal, corresponding to the covariances $\{\hat{\rho}_i\}$, should be synthesized.

7.3 COVARIANCE REALIZATION

In the previous section we have seen that the optimal input design problem can readily be reduced to a standard programming problem by parameterizing the input in terms of $\{\rho_i\}_{i=1}^p$. Since $\rho_1,\ldots,\rho_p$ is an autocorrelation sequence, we must constrain ρ to belong to the following set

$$R = \{\rho \,|\, R_{na+nb-1} > 0;\; R_p \geq 0\} \tag{7.16a}$$

where

$$R_k = \begin{bmatrix} 1 & \rho_1 \ldots & \rho_k \\ \rho_1 & 1 \ldots & \rho_{k-1} \\ \vdots & & \vdots \\ \rho_k & \rho_{k-1} \ldots & 1 \end{bmatrix} \tag{7.16b}$$

Note that it is necessary that $R_{na+nb-1}$ is positive definite. Otherwise the inverse in (7.8) does not exist. This requirement can also be derived from our basic assumptions. We must assume that u(t) is persistently exciting of order na+nb, see Example 4.1. It then follows from (7.10) and Result A1.5 that also $\bar{u}(t)$ is p.e. of order na+nb. This means precisely that $R_{na+nb-1}$ is positive definite.

We will have to distinguish between interior points and boundary points of $\mathcal{R}$. The interior points are given by the set

$$\mathcal{J}(\mathcal{R}) = \{\rho | R_p > 0\} \tag{7.17a}$$

while the boundary points belong to the following sets

$$\mathcal{B}_n(\mathcal{R}) = \{\rho | R_p \geq 0,\ R_{n-1} > 0,\ R_n \text{ singular}\} \tag{7.17b}$$

for some integer n, $na+nb \leq n \leq p$

The constraint $\rho \in \mathcal{R}$ can be efficiently tested using partial autocorrelations. We have for convenience included a discussion on the *partial autocorrelation* (PAC) sequence $\{\phi_k\}_{k=1}^p$ associated to ρ in Appendix A5. This sequence has the following important property, see Lemma A5.4

$$\{R_{n-1} > 0,\ \det(R_n) = 0\}$$

$$\longleftrightarrow \{|\phi_k| < 1 \quad k = 1,\ldots, n-1, |\phi_n| = 1\} \tag{7.18}$$

The partial autocorrelation sequence $\{\phi_k\}_{k=1}^p$ corresponding to a given ρ can be determined for instance using the Levinson-Durbin algorithm (LDA), described in some detail in Appendix A5. This algorithm is given by, see (A5.6)

$$\phi_{k+1} = -a_{k+1,k+1} = (\rho_{k+1}+a_{k,1}\rho_k+\ldots+a_{k,k}\rho_1)/\lambda_k^2$$

$$a_{k+1,i} = a_{k,i}-\phi_{k+1}a_{k,k+1-i} \qquad i = 1,\ldots, k$$

$$\lambda_{k+1}^2 = \lambda_k^2(1-\phi_{k+1}^2) \tag{7.19}$$

with

$$\phi_1 = -a_{1,1} = \rho_1 \qquad \lambda_1^2 = 1-\phi_1^2;\ \ k = 1,2,\ldots$$

As shown in Appendix A5 the above recursion solves iteratively the following linear systems

$$R_{k-1}\begin{bmatrix} a_{k,1} \\ \vdots \\ a_{k,k} \end{bmatrix} = \begin{bmatrix} \rho_1 \\ \vdots \\ \rho_k \end{bmatrix} \qquad k = 1,2,\dots \tag{7.20}$$

$$\lambda_k^2 = 1+a_{k,1}\rho_1+\dots+a_{k,k}\rho_k$$

where $a_{k,1},\dots, a_{k,k},\lambda_k^2$ are the unknowns.

Using (7.18) we can very easily check if ρ belongs to the set of interior points, $\mathcal{J}(\mathcal{R})$, or to the boundary $\mathcal{B}_p(\mathcal{R})$. However, we need some further details to see how the constraint $\rho \in \mathcal{B}_n(\mathcal{R})$, $n < p$ should be tested. Since the optimum may often occur on a boundary $\mathcal{B}_n(\mathcal{R})$ (see e.g., Example 7.1), it is indeed important to analyse such a case. In what follows we will treat interior and boundary points separately in two subsections.

<u>Interior points</u>

Assume that ρ is an interior point, i.e. $\rho \in \mathcal{J}(\mathcal{R})$. Then the recursion (7.19) will give as a byproduct an autoregression (AR)

$$(1+a_{p,1}q^{-1}+\dots+a_{p,p}q^{-p})\bar{u}(t) = \varepsilon(t) \tag{7.21a}$$

with $E\varepsilon(t)\varepsilon(s) = \lambda_p^2\delta_{t,s}$, which exactly matches the autocorrelation sequence $\{\rho_k\}_{k=1}^p$, see Lemma A5.2. Furthermore it is well known that in such a case the polynomial

$$A_p(z) = 1+a_{p,1}z+\dots+a_{p,p}z^p \tag{7.21b}$$

has all its zeros strictly outside the unit disc, see Lemma A5.3.

Hence, whenever the optimal vector $\hat{\rho}$ belongs to $\mathcal{J}(\mathcal{R})$ we can easily realize the optimal input signal as an ARMA process. Just combine (7.10) and (7.21) to get

$$\hat{A}_p(q^{-1})\hat{u}(t) = \hat{\varepsilon}(t) \tag{7.22a}$$

$$\hat{u}(t) = \hat{\sigma}(\hat{\rho})A^*(q^{-1})S(q^{-1})\hat{u}(t) \tag{7.22b}$$

Note that, as anticipated, the optimal input signal cannot in general be obtained without information about the true system parameters. Such an information is required both to evaluate the loss function and in (7.22).

Note also that to generate the optimal input we apparently need initial values for the generating difference equation (7.22a). Fortunately, $\hat{A}_p(z)$ has all zeros strictly outside the unit circle. Then a simple possibility is to start (7.22a) using initial values equal to zero, iterate it a number of times to reach stationarity, and use the sample generated afterwards as optimal input. Needless to say, when some zeros of $\hat{A}_p(z)$ are "close" to the unit circle, the above initialization may become inappropriate. In such a case it may be better to reduce the number of preliminary iterations needed to reach stationarity by using some appropriate initial values to start the recursive calculations implied by (7.22a). It is in fact fairly clear what initial conditions should be used in (7.22a), see the following lemma.

Lemma 7.1. Consider the equation (7.22a), where $\hat{\lambda}_p^2$ and $\{\hat{a}_{p,i}\}$ are uniquely determined through (7.19) (or equivalently (7.20)) from $\hat{\rho} \in \mathcal{J}(R)$. Take $\hat{\hat{u}}(1),\ldots,\hat{\hat{u}}(p+1)$ as a sample drawn out from a (p+1)-variate distribution, independent of $\varepsilon(\cdot)$, with zero mean and covariance matrix $\hat{R}_p$. Generate $\hat{\hat{u}}(p+2)$, $\hat{\hat{u}}(p+3)$ etc with (7.22a). The sequence $\hat{\hat{u}}(k)$, k = 1,2,... so obtained is a realization of a stationary process having the covariance matrix of (p+1) consecutive samples equal to $\hat{R}_p$.

Proof. The proof consists of straightforward calculations.

Let $\hat{\hat{u}}(p+2)$ be generated with (7.22a) using the aforementioned initial values. Then

$$E\hat{\hat{u}}^2(p+2) = E[\hat{a}_{p,1}\ldots\hat{a}_{p,p}]\begin{bmatrix}\hat{\hat{u}}(p+1)\\ \vdots\\ \hat{\hat{u}}(2)\end{bmatrix}[\hat{\hat{u}}(p+1)\ldots\hat{\hat{u}}(2)]\begin{bmatrix}\hat{a}_{p,1}\\ \vdots\\ \hat{a}_{p,p}\end{bmatrix} + \hat{\lambda}_p^2$$

$$= [\hat{a}_{p,1}\ldots\hat{a}_{p,p}]\hat{R}_{p-1}\begin{bmatrix}\hat{a}_{p,1}\\ \vdots\\ \hat{a}_{p,p}\end{bmatrix} + \hat{\lambda}_p^2 = -[\hat{a}_{p,1}\ldots\hat{a}_{p,p}]\begin{bmatrix}\hat{\rho}_1\\ \vdots\\ \hat{\rho}_p\end{bmatrix} + \hat{\lambda}_p^2 = 1$$

The two last equalities follow from (7.20). Furthermore,

$$E\hat{\hat{u}}(p+2)[\hat{\hat{u}}(p+1)\ldots\hat{\hat{u}}(2)] = -[\hat{a}_{p,1}\ldots\hat{a}_{p,p}]\hat{R}_{p-1} = [\hat{\rho}_1\ldots\hat{\rho}_p]$$

The covariance matrix of $[\hat{\hat{u}}(p+2)\ldots\hat{\hat{u}}(2)]^T$ is thus equal to $\hat{R}_p$. Now, the proof can readily be concluded by using induction. ∎

Boundary points

Assume in this subsection that ρ is a boundary point, i.e. $\rho \in B_n(R)$. By analysing this case we shall see how to test for whether a given $\{\rho_k\}$ sequence belongs to $B_n(R)$. We will also examine how inputs corresponding to boundary points should be realized.

We state the result on testing the hypothesis $\rho \in B_n(R)$ as a lemma.

Lemma 7.2. The real numbers $\{\rho_k\}_{k=0}^{p}$, $\rho_0=1$ form an autocorrelation sequence lying in $B_n(R)$ if and only if

$$|\phi_k| < 1 \quad k = 1,\ldots, n-1; \quad |\phi_n| = 1 \tag{7.23a}$$

and $\rho_{n+1},\ldots, \rho_p$ are *uniquely* defined from $\rho_1,\ldots, \rho_n$ through

$$\rho_{n+k} = -a_{n,1}\rho_{n+k-1}-\ldots-a_{n,n}\rho_k \qquad k = 1,2,\ldots \tag{7.23b}$$

where $a_{n,1},\ldots, a_{n,n}$ are given by (7.20).

Proof. Assume first that $\rho \in B_n(R)$. This implies (7.23a) through (7.18).

Let $\bar{u}(t)$ be a process with autocorrelations $\{\rho_k\}$. Since R_n is singular it follows that there exists a linear combination between $\bar{u}(t),\ldots, \bar{u}(t-n)$, for all t. Moreover, since R_{n-1} is nonsingular both u(t) and u(t-n) must appear with non-zero coefficients in this linear combination. Therefore, $\bar{u}(t)$ fulfils a *homogeneous* AR equation of the form

$$\bar{u}(t)+a_{n,1}\bar{u}(t-1)+\ldots+a_{n,n}\bar{u}(t-n) = 0 \tag{7.24}$$

The coefficients $\{a_{n,i}\}$ are uniquely defined from $\rho_1,\ldots, \rho_n$ through (7.20) or equivalently (7.19) (recall that $R_{n-1} > 0$). Now (7.24) implies easily (7.23b), and hence (7.23a), (7.23b) are necessary conditions for $\rho \in B_n(R)$.

Assume next that (7.23a) and (7.23b) hold. Then (7.18) implies that $R_{n-1} > 0$, $R_n \geq 0$, $\det(R_n) = 0$. To prove that $\rho \in B_n(R)$ it is therefore sufficient to show that $\rho_1,\ldots, \rho_p$ is a sequence of autocorrelations. Since $R_n \geq 0$ there exists a signal, say $\bar{u}(t)$, having the autocorrelations at lags 1,..., n equal to $\rho_1,\ldots, \rho_n$. As discussed above $\bar{u}(t)$ will fulfil (7.24) with $\{a_{n,i}\}$ uniquely defined from $\{\rho_i\}_{i=1}^{n}$

through (7.19). It then follows easily that ρ_{n+k}, k = 1,2,... given by (7.23b) are nothing but the autocorrelations of $\bar{u}(t)$ at lags n+1, n+2,... With this observation the sufficiency part of the lemma is also proved. ∎

We proceed now to discuss how a boundary point $\hat{\rho} \in \mathcal{B}_n(R)$ should be realized. It follows from Lemma 7.2 that it is sufficient to realize the first n autocorrelations $\hat{\rho}_1,\ldots,\hat{\rho}_n$ only.

As shown above, the optimal filtered input $\hat{u}(t)$ corresponding to $\hat{\rho} \in \mathcal{B}_n(R)$ will fulfil a homogeneous AR equation of order n (cf. (7.24))

$$\hat{A}_n(q^{-1})\hat{u}(t) = 0 \tag{7.25a}$$

The coefficients of (7.25a) can be obtained with the recursion (7.19). Note from (7.19) that, as expected, we shall get $\hat{\lambda}_n^2 = 0$ (since $\hat{\phi}_n^2 = 1$). Note also that the polynomial $\hat{A}_n(z)$ has all zeros located on the unit circle. This is shown in Lemma A5.4. In the following we will give an alternative proof of this fact. Within this proof we will obtain a characterization of the stationary solutions $\hat{u}(t)$ of (7.25a), that will prove useful for the subsequent analysis.

Consider the following series of obvious equivalences [$\phi_{\hat{u}}(\omega)$ being the spectral density of $\hat{u}(t)$]:

$$\begin{aligned}
(7.25a) &\Longleftrightarrow \oint_{-\pi}^{\pi} |\hat{A}_n(e^{i\omega})|^2 \phi_{\hat{u}}(\omega)d\omega = 0 \\
&\Longleftrightarrow |\hat{A}_n(e^{i\omega})|^2 \phi_{\hat{u}}(\omega) \equiv 0 \qquad (7.25b) \\
&\Longleftrightarrow \{\phi_{\hat{u}}(\omega) > 0 \Rightarrow \hat{A}_n(e^{i\omega}) = 0\}
\end{aligned}$$

Since $\hat{\rho} \in \mathcal{B}_n(R)$, $\hat{u}(t)$ is a persistently exciting process of order n so that its spectrum is non-zero in precisely n distinct points, see e.g. Result A1.3. In view of (7.25b), $\hat{A}_n(z)$ must then have n (i.e. all) zeros on the unit circle.

The above discussion, see (7.25b), also shows that equation (7.25a) will have infinitely many solutions with different autocorrelations. Now, it can easily be seen that a result similar to Lemma 7.1 can be proved for the homogeneous AR equation (7.25a) as well. It might then be expected that by initializing (7.25a) with a sample $[\hat{u}(1)\ldots\hat{u}(n+1)]^T$ drawn out from a distribution with covariance matrix $\hat{R}_n$, we might get the desired solution $\hat{u}(t)$. Unfortunately this is not true, since $\hat{u}(t)$ generated in that way will not be an ergodic process! Equation (7.25a) must thus be used in another way to generate $\hat{\hat{u}}(t)$. The input $\hat{\hat{u}}(t)$ corresponding to $\hat{\rho} \in \mathcal{B}_n(R)$ will have a discrete spectrum containing exactly n distinct frequencies, cf Result A1.3. Furthermore, it follows from (7.25b) that the spectral frequencies

$$0 \le \hat{\omega}_1 < \hat{\omega}_2 < \ldots < \hat{\omega}_m \le \pi \tag{7.26a}$$

are uniquely specified by

$$\hat{A}_n(e^{i\hat{\omega}_k}) = 0 \qquad k = 1,\ldots, m \tag{7.26b}$$

where m is given by

$$m = \begin{cases} (n+1)/2 & \text{if } \hat{\omega}_1 = 0 \text{ or } \hat{\omega}_m = \pi\text{, but not both} \quad (n \text{ odd}) \\ (n+2)/2 & \text{if } \hat{\omega}_1 = 0 \text{ and } \hat{\omega}_m = \pi \\ & \qquad\qquad (n \text{ even}) \\ n/2 & \text{if } \hat{\omega}_1 \neq 0 \text{ and } \hat{\omega}_m \neq \pi \end{cases} \tag{7.26c}$$

Thus, we can easily find $\{\hat{\omega}_k\}$ by solving the equation $\hat{A}_n(z) = 0$.

Once the frequencies have been determined, the associated spectral powers, say $\{\hat{\beta}_k > 0\}_{k=1}^m$, can be determined from the following system of linear equations.

$$1 = \sum_{j=1}^{m} \hat{\beta}_j \tag{7.26d}$$

$$\hat{\rho}_k = \sum_{j=1}^{m} \hat{\beta}_j \cos(\hat{\omega}_j k) \qquad k = 1,2,\ldots, n$$

Recall that $1,\hat{\rho}_1,\ldots, \hat{\rho}_n$ determine the whole covariance sequence of $\hat{\hat{u}}(t)$, cf Lemma 7.2. Then, since the spectral density associated with a given covariance function is unique, it follows that (7.26d) will have a unique solution with respect to $\{\hat{\beta}_j\}_{j=1}^m$. Furthermore, the solution can be obtained from the first m equations of (7.26d)

$$\begin{bmatrix} 1 & 1 & \ldots & 1 \\ \cos\hat{\omega}_1 & \cos\hat{\omega}_2 & \ldots & \cos\hat{\omega}_m \\ \vdots & \vdots & & \vdots \\ \cos(m-1)\hat{\omega}_1 & \cos(m-1)\hat{\omega}_2 & \ldots & \cos(m-1)\hat{\omega}_m \end{bmatrix} \begin{bmatrix} \hat{\beta}_1 \\ \hat{\beta}_2 \\ \vdots \\ \hat{\beta}_m \end{bmatrix} = \begin{bmatrix} 1 \\ \hat{\rho}_1 \\ \vdots \\ \hat{\rho}_{m-1} \end{bmatrix} \tag{7.26e}$$

since the matrix appearing in (7.26e) is nonsingular, see Lemma A3.10.

To summarize, the spectral frequencies of $\hat{u}(t)$ are obtained from (7.26b), and their associated weights from (7.26e). In the time domain the optimal filtered input $\hat{u}(t)$ can then be synthesized as

$$\hat{u}(t) = \sum_{i=1}^{m} \hat{\gamma}_i \cos\hat{\omega}_i t \tag{7.27a}$$

where

$$\hat{\gamma}_i = \begin{cases} \sqrt{2\hat{\beta}_i} & \text{if } 0 < \hat{\omega}_i < \pi \\ \sqrt{\hat{\beta}_i} & \text{if } \hat{\omega}_i = 0 \text{ or } \hat{\omega}_i = \pi \end{cases} \tag{7.27b}$$

Once $\hat{u}(t)$, t = 1,2,... is available, the optimal input can easily be determined from (7.22b).

Alternatively, notice from (7.22b) that $\hat{u}(t)$ will have the same optimal frequencies $\hat{\omega}_1, \ldots, \hat{\omega}_m$ as $\hat{u}(t)$, but the corresponding weights are in the case of $\hat{u}(\bullet)$ given by

$$\hat{\beta}_k^* = \hat{\sigma}^2(\hat{\rho})|A^*(e^{i\hat{\omega}_k})S(e^{i\hat{\omega}_k})|^2\hat{\beta}_k \quad k = 1,\ldots, m$$

Hence $\hat{u}(t)$ can directly be synthesized as a sum of suitably chosen sinusoids, and this appears to be a computationally cheaper way.

Finally, note that the optimal input signal (7.27), (7.22b) will have a nonzero mean whenever $\hat{\omega}_1 = 0$. The constraint $\hat{u}(t) \in \mathcal{D}$ is, though, fulfilled. Then, a possible nonzero mean of the input should not be a problem in practice.

7.4 ALGORITHMIC ASPECTS

There are at least two ways to organise the search procedure involved by the optimal input design problem previously discussed.

One way is to conduct the optimization directly with respect to ρ. Whenever the optimal values $\hat{\rho}$ are situated well inside $\mathcal{J}(R)$, this way to proceed should not give any difficulty. However, it often happens that the optimal point is located on the boundary $\mathcal{B}(R)$. Note that the equations defining the boundary are very complicated functions of ρ. An optimization with respect to ρ will therefore require a sophisticated algorithm in order to handle the possible boundary case.

A second way to proceed, which we shall discuss in this section, is intended to overcome the aforementioned difficulty.

Let us first reparameterize the (filtered) input $\ddot{u}(t)$ in terms of the PAC sequence $\{\phi_k\}$. Then constrain ϕ_k to be in the interval (-1,1), e.g. by the transformation (Jones (1980))

$$\phi_k = \frac{1-e^{\alpha_k}}{1+e^{\alpha_k}} \qquad k = 1,\dots, p \tag{7.28}$$

The nonlinear optimization will now be carried out (without constraints) with respect to $\{\alpha_k\}_{k=1}^{p}$. An initial point for the search could be obtained by performing a random search within the cube $|\phi_k| \le 1$, $k = 1,\dots, p$, or it can simply be taken as $\{\phi_k\} = 0$. Since the dependence of the loss function on $\{\alpha_k\}$ appears to be quite complicated (see also (7.29) below), the search should be continued using an optimization algorithm that needs only function evaluations. The sequence $\{\rho_k\}$ required to evaluate the loss function can be calculated from $\{\phi_k\}$ by rearranging the LDA (7.19):

$$\begin{aligned}
\rho_{k+1} &= \lambda_k^2\phi_{k+1}-a_{k,1}\rho_k-\dots-a_{k,k}\rho_1 \\
\lambda_{k+1}^2 &= \lambda_k^2(1-\phi_{k+1}^2) \\
a_{k+1,k+1} &= -\phi_{k+1} \\
a_{k+1,i} &= a_{k,i}-\phi_{k+1}a_{k,k+1-i} \qquad i = 1,\dots, k
\end{aligned} \tag{7.29a}$$

The initial values for the recursion (7.29a) are the following

$$\rho_1 = -a_{1,1} = \phi_1 \qquad \lambda_1^2 = 1-\phi_1^2 \tag{7.29b}$$

The sequence $\{\rho_k\}_{k=1}^{p}$ obtained from (7.28), (7.29) will clearly be positive (semi) definite, as required; that is, it will belong to R. This is obvious in the case $|\phi_k| < 1$, $k = 1,\dots, p$ when there is a one-to-one map between $\{\phi_k\}_{k=1}^{p}$ and $\{\rho_k\}_{k=1}^{p}$. The limiting case when $|\phi_k| = 1$ for some k, needs additional comments.

Let us assume that $\rho \in B_n(R)$. In the ϕ-parameter space, the corresponding values $\phi_1,\dots, \phi_n$ are uniquely defined and fulfil

$$|\phi_k| < 1 \quad k = 1,\ldots n-1 \quad |\phi_n| = 1 \tag{7.30}$$

while $\phi_{n+1},\ldots, \phi_p$ are not defined (cf. the discussion in Section 7.3). Conversely, any sequence $\phi_1,\ldots, \phi_p$ where the first n elements fulfil (7.30), and the remaining ones are *arbitrary* will through (7.29) define a unique $\rho \in B_n(R)$. That this is true, is quite clear at least if we use the recursion (8.29) in the following manner. All equations are iterated until $k = n-1$ and then $\rho_{n+1},\ldots,\rho_p$ are computed from (7.23b).

The above result remains true even if we iterate (7.29) until $k = p-1$. To show this, notice first that it can easily be seen from the last two equations of (7.29a) that

$$A_{k+1}(z) = A_k(z)-z^{k+1}\phi_{k+1}A_k(z^{-1}) \tag{7.31}$$

Due to (7.30) and Lemma A5.4, $A_n(z)$ will have all zeros on the unit circle. It then follows from (7.31) that any zero of $A_n(z)$ will also be a zero of $A_k(z)$, $k > n$. Let $\bar{u}(t)$ be the process with autocorrelations $\rho_1,\ldots, \rho_n$ obtained from (7.29) and the remaining ones given by (7.23b). It is known to fulfil

$$A_n(q^{-1})\bar{u}(t) = 0$$

see (7.24). According to the above discussion and (7.25b), it will also fulfil

$$A_k(q^{-1})\bar{u}(t) = 0 \qquad k \geq n$$

which implies that

$$A_k(q^{-1})\rho_{k+1} = 0 \qquad k \geq n$$

This is simply the first equation of (7.29a) (since $\lambda_k^2 = 0$ for $k \geq n$). The above discussion also shows that the recursion (7.29), unlike (7.19), does not need to be stopped at iteration n when $\rho \in B_n(R)$ or equivalently $\{\phi_k\}$ fulfils (7.30). Needless to say, in practice it is advisable - in order to save computation time - to detect the first ϕ_k, say ϕ_n, with unit magnitude and to stop (7.29) at iteration n (generating the remaining correlations ρ_{n+1}, ρ_{n+2} etc with (7.23b) as already discussed).

Note that in practice we cannot have exactly $|\phi_k| = 1$, see (7.28). However, for reasonably high values of $|\alpha_k|$, $|\phi_k|$ will be so close to 1 that we can disregard this point as being minor.

The important point which should be retained from the above discussion, is the arbitrariness of $\phi_{n+1}, \ldots, \phi_p$. It has the following consequences for the optimization procedure, whenever $\hat{\rho} \in \mathcal{B}_n(\mathcal{R})$.

- Since $\phi_{n+1}, \phi_{n+2}, \ldots$ cannot be expected to converge in the course of optimization, the stopping condition should be based on the relative decrease of the loss from one iteration to another, rather than on the relative changes of $\{\phi_k\}$ (or $\{\alpha_k\}$). Also, it may not be advisable to use Hessian-based search procedures, since the Hessian matrix is expected to become singular close to the area of nonisolated minimum points. Such difficulties may however be avoided by detecting the first ϕ_k, say ϕ_n, with magnitude that tends to 1 in the course of optimization *and* conducting the optimization further only with respect to $\phi_1, \ldots, \phi_n$.

- Assume that $\hat{\rho} \in \mathcal{B}_n(\mathcal{R})$ and that an upper bound $\bar{n}$ on n, which is less than p $(n \leq \bar{n} < p)$ is known. Then we can clearly reduce the dimension of the optimization problem from p to $\bar{n}$ by fixing $\alpha_{\bar{n}+1}, \ldots, \alpha_p$ to some given values, e.g. zero.

  It is for instance known that for the problem under discussion there always exists an optimal input comprising not more than 2(na+nb) distinct frequencies, see e.g. Goodwin and Payne (1977). For such an input process $R_{2(na+nb)}$ will be singular cf e.g. Result A1.3, so if

  $$p > 2(na+nb)$$

  we can take $\bar{n} = 2(na+nb)$.

Now, there are a number of questions concerning the application of the above optimal input design procedure, which have not yet been treated. For instance:

- What is the sensitivity of the parameter accuracy, obtained when using the "optimal" input, to uncertainties in the initial parameter estimates required to apply the design procedure?

- How much is gained in accuracy by using an optimal input as compared to, say, a white input process?

In order to give some answers to the above questions we will resort to Monte-Carlo simulations.

Example 7.1 A D-optimal input design

Consider the following second-order system

$$(1+a_1q^{-1}+a_2q^{-2})y(t) = bq^{-1}u(t)+\frac{1}{1+dq^{-1}}e(t) \qquad (7.32)$$

The multistep algorithm will be used to get estimates of the parameters $[a_1\ a_2\ b\ d]^T$. State the optimal input design problem as

$$\min_{u(t)\,\in\,\mathcal{D}'_y} \det(P_4) \qquad (7.33)$$

with $\mathcal{D}'_y$ defined by (7.4b) with $\sigma_x^2 = 1$ and P_4 by (6.51b). Now, it follows from (6.52) that

$$\det(P_4) = \det(P_{IV}^{opt})\det(Q_{\beta\beta}^{-1})$$

Since the input does not affect $Q_{\beta\beta}$ we conclude that (7.33) is equivalent to

$$\min_{u(t)\,\in\,\mathcal{D}'_y} \det(P_{IV}^{opt}) \qquad (7.34)$$

with P_{IV}^{opt} given by (7.6) where for the present case $H^{-1}(q^{-1}) = 1+dq^{-1}$. Further, using the parameterization (7.9), (7.11) of the input, (7.34) is reduced to the following standard nonlinear programming problem (cf. (7.8)):

$$\max_{\rho=\begin{bmatrix}\rho_1\\ \rho_2\\ \rho_3\end{bmatrix}\in R} \hat{\sigma}^6(\rho)\cdot\det\begin{bmatrix}1+d^2+2d\rho_1 & d+(1+d^2)\rho_1+d\rho_2 & d\rho_1+(1+d^2)\rho_2+d\rho_3\\ d+(1+d^2)\rho_1+d\rho_2 & 1+d^2+2d\rho_1 & d+(1+d^2)\rho_1+d\rho_2\\ d\rho_1+(1+d^2)\rho_2+d\rho_3 & d+(1+d^2)\rho_1+d\rho_2 & 1+d^2+2d\rho_1\end{bmatrix} \qquad (7.35)$$

where $\hat{\sigma}^2(\rho)$ is a constant given by (cf. 7.13))

$$\hat{\sigma}^2 = \frac{1}{b^2} \qquad (7.36)$$

The optimization problem (7.35) is then transformed into one without constraints by using (7.28), (7.29). The optimization with respect to $\{\alpha_k\}$ is performed using a variable metric algorithm, Fletcher (1971), initialized with $\alpha_k = 0$, k = 1,2,3.

The initial parameter estimates needed to apply the optimal design procedure are obtained as described in the following scheme which also summarizes the whole Monte-Carlo simulation procedure:

i) Generate 500 data pairs with (7.32), using a white input process $u(t) \in \mathcal{D}_y'$. Choose $Ee^2(t)$ such that the signal-to-noise ratio S have a prescribed value.

ii) Apply the multistep algorithm to the data generated in step i), to get initial parameter estimates.

iii) Repeat steps i)-ii) for 50 different realizations of the noise process used in step i). Evaluate the sample covariance matrix of the estimates obtained in step ii).

iv) Choose one of the estimates obtained in step iii). Determine the "optimal" input using this estimate instead of the true parameter values.

v) Generate with (7.32) a new set of 500 data pairs using the optimal input and the same noise sequence as employed in step i).

vi) Apply the multistep algorithm to the new data set to get final parameter estimates.

vii) Similarly with step iii), but now for steps v)-vi).

viii) Choose other three initial estimates from those obtained in step iii). Repeat steps iv)-vii) for each of these estimates in part.

ix) Finally, repeat steps iv)-vii) using the true parameter values to design the optimal input (in step iv).

Note that steps viii)-ix) are included to show, to some extent, the sensitivity of the "optimal" accuracy to uncertainties in the initial parameter estimates used to design the "optimal" input.

The results obtained by applying the above procedure for several values of the system parameters are shown in Tables 7.1-7.3. Some comments on these results follow.

System			a_1	a_2	b	d	S
S_1	True values		-1.000	0.600	1.000	0.500	3.3
	Initial estimates	init. 0	-1.000	0.600	1.000	0.500	-
		init. 1	-1.006	0.619	0.962	0.567	-
		init. 2	-1.001	0.607	0.998	0.524	-
		init. 3	-1.005	0.608	0.993	0.460	-
		init. 4	-0.981	0.585	0.986	0.481	-
S_2	True values		1.000	0.500	1.000	-0.900	3.4
	Initial estimates	init. 0	1.000	0.500	1.000	-0.900	-
		init. 1	0.996	0.487	1.003	-0.881	-
		init. 2	1.004	0.500	0.999	-0.920	-
		init. 3	1.016	0.507	0.982	-0.853	-
		init. 4	0.994	0.499	0.992	-0.909	-
S_3	True values		0.600	0.800	1.000	-0.700	2.6
	Initial estimates	init. 0	0.600	0.800	1.000	-0.700	-
		init. 1	0.603	0.808	0.993	-0.672	-
		init. 2	0.614	0.804	1.005	-0.647	-
		init. 3	0.599	0.812	1.006	-0.721	-
		init. 4	0.606	0.798	1.008	-0.744	-
S_4	True values		-1.500	0.700	1.000	0.800	1.9
	Initial estimates	init. 0	-1.500	0.700	1.000	0.800	-
		init. 1	-1.499	0.702	1.009	0.768	-
		init. 2	-1.494	0.686	1.023	0.730	-
		init. 3	-1.463	0.677	1.020	0.820	-
		init. 4	-1.489	0.687	1.017	0.811	-

Table 7.1. The system parameter values used in the Monte-Carlo experiment. The (five) initial parameter estimates used in each case to design the "optimal" input(s) are also shown.

System	Initialization number	Optimal input parameters			
		ω_1	ω_2	γ_1	γ_2
S_1	0	0	1.647	0.346	1.375
	1	0	1.622	0.367	1.384
	2	0	1.638	0.350	1.358
	3	0	1.666	0.351	1.420
	4	0	1.658	0.352	1.400
S_2	0	1.569	π	1.293	0.289
	1	1.566	π	1.296	0.283
	2	1.570	π	1.297	0.287
	3	1.566	π	1.336	0.289
	4	1.569	π	1.298	0.294
S_3	0	1.549	π	0.774	0.693
	1	1.546	π	0.786	0.700
	2	1.540	π	0.802	0.683
	3	1.554	π	0.755	0.697
	4	1.557	π	0.760	0.684
S_4	0	0	1.579	0.115	1.782
	1	0	1.582	0.116	1.771
	2	0	1.587	0.109	1.754
	3	0	1.576	0.121	1.704
	4	0	1.578	0.113	1.741

Table 7.2. The parameters of the optimal inputs $\hat{u}(t) = \gamma_1 \cos \omega_1 t + \gamma_2 \cos \omega_2 t$ [$\{\omega_i\}$ are given in rad.]

System			a_1	a_2	b	d	Accuracy criterion $10^{16}\det\hat{P}_4$
			Parameters				
S_1	True values		-1.000	0.600	1.000	0.500	-
	Estimated values	White input	-0.999±0.010	0.599±0.011	0.997±0.013	0.490±0.038	15.69
		Optimal input 0	-1.001±0.010	0.600±0.009	1.000±0.009	0.491±0.038	7.11
		1	-1.001±0.012	0.600±0.010	0.999±0.009	0.492±0.037	8.07
		2	-1.002±0.011	0.601±0.011	1.000±0.010	0.492±0.037	8.64
		3	-1.003±0.011	0.601±0.008	0.998±0.011	0.492±0.038	6.86
		4	-1.001±0.009	0.601±0.011	1.002±0.010	0.491±0.038	7.67
S_2	True values		1.000	0.500	1.000	-0.900	-
	Estimated values	White input	0.999±0.010	0.498±0.010	1.002±0.012	-0.891±0.021	1.69
		Optimal input 0	1.001±0.008	0.502±0.008	1.000±0.010	-0.892±0.021	1.56
		1	1.001±0.009	0.501±0.010	0.999±0.010	-0.892±0.021	1.54
		2	1.001±0.009	0.503±0.008	1.000±0.009	-0.892±0.021	1.56
		3	1.001±0.009	0.501±0.009	0.999±0.010	-0.892±0.021	1.31
		4	1.001±0.009	0.502±0.008	1.000±0.010	-0.892±0.021	1.48
S_3	True values		0.600	0.800	1.000	-0.700	-
	Estimated values	White input	0.600±0.011	0.801±0.011	1.005±0.020	-0.699±0.029	32.20
		Optimal input 0	0.602±0.010	0.800±0.014	0.999±0.011	-0.699±0.029	11.44
		1	0.600±0.010	0.798±0.011	0.998±0.012	-0.698±0.029	9.36
		2	0.599±0.010	0.799±0.010	1.001±0.010	-0.698±0.030	5.79
		3	0.601±0.012	0.801±0.012	1.000±0.010	-0.699±0.029	10.92
		4	0.600±0.012	0.800±0.012	1.000±0.011	-0.699±0.029	9.59
S_4	True values		-1.500	0.700	1.000	0.800	-
	Estimated values	White input	-1.489±0.018	0.691±0.020	1.012±0.030	0.789±0.030	193.19
		Optimal input 0	-1.495±0.013	0.697±0.011	0.996±0.010	0.789±0.029	6.90
		1	-1.495±0.014	0.697±0.012	1.000±0.011	0.789±0.029	8.98
		2	-1.498±0.015	0.701±0.009	1.001±0.013	0.790±0.029	9.96
		3	-1.498±0.013	0.700±0.009	0.997±0.011	0.790±0.029	6.74
		4	-1.496±0.013	0.698±0.011	0.996±0.011	0.789±0.029	7.61

Table 7.3. Estimated parameters (means and standard deviations computed from 50 realizations) and accuracy criteria values corresponding to a white input process as well as to the "D-optimal" input(s).

- Note from Table 7.1 that in each case "init. 0" denotes the use of the true parameter values as initial estimates. Note also from this table that the other initial parameter estimates used are quite scattered around the true values.
- In all cases the optimal autocorrelations of the input were found to be located on the boundary $B_3(R)$, see Table 7.2. Typically, the used search procedure needed about 100 loss function evaluations to reach the "optimum" (recall that the starting point was in all cases equal to zero). At the end of the search, $|\phi_3|$ was frequently in the interval [0.9999, 1].
- It can be seen from Table 7.2 that the optimum input parameters are not too sensitive to uncertainties in the initial information required to apply the design procedure. The sensitivity of the "optimal" accuracy to these uncertainties seems also not to be too important, see Table 7.3 (note that the variation of $\det(\hat{P}_4)$ in Table 7.3 is partially due to sample fluctuations).
- In all cases, the optimal input leads to better accuracy (i.e., to smaller $\det(\hat{P}_4)$) than the white input signal, see Table 7.3. As expected, the gain of accuracy depends on the system parameters (including the signal-to-noise ratio). In some cases the gain is minor (as, e.g. for system S_2); in others it is substantial (particularly for S_4).
- Finally, it is interesting to notice from Table 7.3 that without exception the optimal input leads also to smaller $\mathrm{tr}(\hat{P}_4)$ than the white input process (recall that the optimal input minimizes $\det(P_4)$ not $\mathrm{tr}(P_4)$). ■

7.5 REMARKS AND BIBLIOGRAPHICAL NOTES

The basic idea behind the optimal input design procedure presented in this chapter was to transform the given optimal input design problem into a standard nonlinear programming one, by introducing a suitable parameterization of the input. The specific parameterization through $\{\rho_i\}$ [or $\{\phi_i\}$] discussed here appears in recent works by the authors, see Söderström and Stoica (1979 a), Stoica and Söderström (1982 f), but the computational aspects involved were not penetrated in detail before.

Other possible parameterizations of the input which can be used to design the optimal input signals both for PEM and for optimal IVM appear in Mehra (1974, 1976, 1981), Goodwin and Payne (1977), Zarrop (1979 a,b) etc.. The parameterization of the input considered in these references is essentially of the following form

$$u_\rho(t) = \sum_{i=1}^{\bar{p}} \gamma_i \cos \omega_i t \qquad (7.37)$$

Note the similarity with (7.27). One can prove, see Mehra (1974), Goodwin and Payne (1977) that if $\bar{p}$ = na+nb, then (7.37) fulfils (7.5). Using the parameterization (7.37) we can thus find the optimal input by solving a 2(na+nb)-dimensional optimization problem. Note that our approach leads to an optimization problem of a dimension which in general is smaller than 2(na+nb). Indeed for values of na, nb, ns and nt encountered in practice, we may expect that p - as defined by (7.14) - will often be smaller than 2(na+nb). When this is not the case, our p-dimensional problem, can, however, be reduced to one of dimension 2(na+nb), as shown in Section 7.4.

Some refinements of the approach based on (7.37) have been discussed by Zarrop (1979 a,b), who showed that under certain (sufficient) conditions on the system degrees it is feasible to use in (7.37) the minimally possible number of sinusoids, that is $\bar{p}$ = [(na+nb)/2]. It appears that under those conditions, requiring e.g. that nt = 0, p will in general be equal to na+nb-1. Similarly as above, in cases with p > na+nb (under the conditions of Zarrop (1979 a, b)), we can use only $\phi_1, \ldots, \phi_{na+nb}$ as independent variables in the optimization.

Finally, note that the covariance matrix (and hence the loss function) corresponding to a given input of the form (7.37) can conveniently be calculated by using frequency-domain techniques, see e.g. Goodwin and Payne (1977).

PART III

PRACTICAL ASPECTS AND CASE STUDIES

Chapter 8

SUMMARY OF THE ANALYSIS AND HINTS FOR THE USER

In this chapter we will summarize what we have derived in the two previous parts of the book. We will stress the main ideas while technical details will not always be repeated. The implication of the theoretical results for practical applications will be discussed in particular. Thus also some hints for the potential user of IVMs will be given. In the next chapter some case studies are presented, which also illustrate some of the practical hints to be presented here.

8.1 BASIC FACTS ON IV ESTIMATION

In Chapter 2 we introduced the concepts *system* and *model structure*. We also made some *basic assumptions*. For convenience these will be briefly repeated here.

Let y(t) be the ny-dimensional output at time t, and u(t) the nu-dimensional input of a discrete-time, asymptotically stable, stochastic system denoted by S. Consider the following vector difference model of S, cf (2.1)

$$M:\quad A(q^{-1},\theta)y(t) = B(q^{-1},\theta)u(t)+\varepsilon(t) \tag{8.1}$$

where θ is a $n\theta$-dimensional vector of unknown parameters, $\{\varepsilon(t)\}$ are the residuals, and $A(q^{-1},\theta)$, $B(q^{-1},\theta)$ are the following polynomials in the unit delay operator q^{-1}

$$\begin{aligned} A(q^{-1},\theta) &= I+A_1(\theta)q^{-1}+\ldots+A_{na}(\theta)q^{-na} \\ B(q^{-1},\theta) &= \phantom{I+{}} B_1(\theta)q^{-1}+\ldots+B_{nb}(\theta)q^{-nb} \end{aligned} \tag{8.2}$$

There are several ways to parameterize the model (8.1), i.e. to let the matrices $A_1(\theta),\ldots, B_{nb}(\theta)$ depend on the parameter vector. We have previously given some possibilities, see Examples 2.1-2.5. We will have more to say on the model structure selection in Section 8.4.

Some basic assumptions on S, M and the *experimental condition* X, were introduced in Chapter 2. They will now be repeated for easy reference. We assume that all these assumptions are generally valid.

A1	The matrix coefficients $\{A_i(\theta)\}$, $\{B_i(\theta)\}$ are linear functions of θ.
A2	The system S is linear, of finite order and asymptotically stable. The output $y(t)$ fulfils $$y(t) = x(t)+w(t) \qquad x(t) = G(q^{-1})u(t)$$ where $x(t)$ is the noise-free output, $G(q^{-1})$ the matrix transfer function, and $w(t)$ an additive disturbance
A3	$w(t)$ is a full rank stationary stochastic process with zero mean and rational spectral density matrix
A4,A5	There exists a unique vector θ^* such that $A(z,\theta^*)^{-1}\ B(z,\theta^*) \equiv G(z)$
A6	The input $u(t)$ is a persistently exciting signal that is ergodic with respect to second order moments
A7	The input $u(t)$ and the disturbance $w(s)$ are independent for all t and s

It follows from these assumptions that the model (8.1) can be written as

$$M: \quad y(t) = \phi^T(t)\theta+\varepsilon(t) \tag{8.3}$$

The matrix $\phi(t)$ is of dimension $n\theta|ny$. Its entries are linear combinations of delayed input and output components. Similarly the system S will fulfil, cf Assumptions A2-A5

$$S: \quad y(t) = \phi^T(t)\theta^*+H(q^{-1})e(t) \tag{8.4}$$

where $e(t)$ is zero mean white noise with a positive definite covariance matrix Λ. The filter $H(q^{-1})$ and its inverse are asymptotically stable. We call θ^* the true parameter vector.

Most of the assumptions are mild. Some of them, though, need further comments. Assumption A2 on the true system seems reasonable from a *theoretical* point of view. Indeed, any linear finite order stochastic system can be represented by a (vector) difference equation of the form (8.1), (8.4), at least after a preliminary shifting of the input. Furthermore, certain types of nonlinear systems (e.g. Hammerstein systems, see Example 2.3) can be described by equations of the form (8.4). From a *practical* point of view, A2 is however restrictive, since real systems are hardly exactly linear and of finite order. Nevertheless, A2 is a crucial assumption in the analysis (without A2 satisfied it would e.g. not be possible to speak about consistency of IV estimates). It should also be noted that in many cases real systems can be well approximated with linear finite order models (cf., for instance, the numerical applications to real data in the next chapter). Note that it is a consequence of A2, A4, A5 that

$$S \in M \tag{8.5}$$

i.e. the system belongs to the model structure. We will discuss later in Section 8.4 how to choose appropriate model structures when (8.5) is unlikely to hold. Such a situation is quite realistic in practice. Then it is relevant to choose M so that the estimated model gives a good *approximation* (in a given sense) of the system.

It is worth noting here that in practice it may often be advisable to use "pseudo-canonical" rather than canonical parameterizations, see e.g. Hannan (1969, 1971, 1975, 1976), Kashyap and Rao (1976), Stoica (1982 b). For a pseudo-canonical parameterization of the system dynamics, the assumptions A4, A5 will be fulfilled only *generically* [that is, there will be systems which do not satisfy A4, A5, but in the θ-parameter space such systems belong to some set of lower dimension]. As compared to a canonical parameterization, a pseudo-canonical one has hence the drawback that its use makes impossible the consistent estimation of some (but few) systems. However, this should not be viewed as a major problem since such "pathological" systems are unlikely to be encountered in practice. Furthermore, the number of structural indices required to define a pseudo-canonical model is in general considerably smaller than that required to describe various canonical parameterizations. Consider, for instance, the "full polynomial form" described in Example 2.5. It is a pseudo-canonical parameterization, see Lemma 2.1 and the related discussion. It requires two integer-valued parameters to be specified, namely the orders na, nb. A canonical parameterization needs usually ny+1 integer values. In practice, the structural indices will have to be scanned. This operation is clearly time consuming and as a consequence the smaller number of integer parameters needed to define a pseudo-canonical parameterization (e.g. the "full polynomial form") could be an important practical advantage for many applications.

The assumption A7 means from a practical point of view that the system operates in open loop. This assumption has been used throughout the analysis. The construction of several IVMs is also based on A7. However, as we have occasionally mentioned in Section 3.4, there also exist IV variants which allow closed loop operation. We also remarked in Section 5.3 that the derivation of the asymptotic distribution of the

IV parameter estimates can be extended to allow some dependence between $u(\cdot)$ and $w(\cdot)$.

In Chapter 3 we discussed basic and extended IVMs for the estimation of the "true" parameter vector θ^* describing the system dynamics. The idea behind the IVMs is simply to get a consistent estimate of θ^* in an easy way. Indeed, let $F(q^{-1})$ be an asymptotically stable, ny|ny-dimensional filter and assume that a nz|ny-matrix $Z(t)$, $nz \geq n\theta$, can be found such that, cf (3.27), (3.28)

$$\text{rank } R = n\theta \qquad R \triangleq EZ(t)\cdot F(q^{-1})\phi^T(t) \tag{8.6}$$

$$EZ(t)\cdot F(q^{-1})v(t) = 0 \qquad (v(t) = H(q^{-1})e(t)) \tag{8.7}$$

Then it can easily be seen from (8.4) that θ^* is the unique solution of the following linear system of equations

$$[EZ(t)\cdot F(q^{-1})\phi^T(t)]\theta = EZ(t)\cdot F(q^{-1})y(t) \tag{8.8}$$

Under some regularity assumptions (e.g. concerning the ergodicity of the second order moments involved in (8.8)) it is thus possible to get a consistent estimate $\hat{\theta}$ of θ^* by solving the following system of linear equations

$$[\sum_{t=1}^{N} Z(t)\cdot F(q^{-1})\phi^T(t)]\hat{\theta} = [\sum_{t=1}^{N} Z(t)\cdot F(q^{-1})y(t)] \tag{8.9}$$

where N is the number of data.

For $nz = n\theta$ and $F(q^{-1}) \equiv I$ the estimate (8.9) reduces to the well-known *basic* IV estimate

$$\hat{\theta} = [\sum_{t=1}^{N} Z(t)\phi^T(t)]^{-1}[\sum_{t=1}^{N} Z(t)y(t)] \tag{8.10}$$

where the inverse exists, at least for N large enough, in view of the assumption (8.6).

For $nz > n\theta$ and/or $F(q^{-1}) \not\equiv I$ the estimate (8.9) can be viewed as an extension of (8.10). For this reason we have called (8.9) an *extended* IV estimate. Note that for $nz > n\theta$, the system of equations (8.9) is overdetermined. It is then to be solved in a least-squares sense, for example using the orthogonalization method, Stewart (1973). The estimate will then be defined by

$$[\sum_{t=1}^{N} Z(t)\cdot F(q^{-1})\phi^T(t)]^T Q[\sum_{t=1}^{N} Z(t)\cdot F(q^{-1})\phi^T(t)]\hat{\theta}$$

$$= [\sum_{t=1}^{N} Z(t)\cdot F(q^{-1})\phi^T(t)]^T Q[\sum_{t=1}^{N} Z(t)\cdot F(q^{-1})y(t)] \qquad (8.11)$$

where Q, a positive definite weighting matrix, was introduced as a further generalization of (8.9).

Needless to say, the extended IV estimate (8.9) will require more computations than (8.10). The need for the extensions must therefore be clearly motivated. Two reasons to use an extended IVM are as follows:

- It may be easier to fulfil the consistency conditon (8.6) if nz is taken larger than nθ.

- The additional "degrees of freedom" obtained with $F(q^{-1})$ and an augmented IV matrix can potentially give an increased accuracy of the parameter estimate.

We have previously demonstrated, see e.g. Example 3.5 and Theorem 6.1 respectively, that these reasons really are applicable and make sense in practice.

Return now to the consistency conditions (8.6), (8.7). Usually the IV matrix Z(t) is constructed using lagged and possibly filtered input components. According to A7 the second consistency condition is then trivially fulfilled. Furthermore, (8.6) becomes

$$\text{rank } R = n\theta \qquad R = EZ(t)\cdot F(q^{-1})\tilde{\phi}^T(t)$$

where $\tilde{\phi}(t)$ is the noise-free part of $\phi(t)$ (i.e. $\tilde{\phi}(t)$ is obtained from $\phi(t)$ by replacing everywhere y(t) by x(t)). It is not difficult to show (see Section 4.1) that for (8.12) to hold it is *necessary* that u(t) is a persistently exciting signal of a sufficiently high order, and A4, A5 are satisfied.

In Chapter 4 we have also derived *sufficient* conditions for the consistency requirement (8.6) to be fulfilled. We have proved that it is possible to get consistent estimates under mild conditions. Moreover, the analysis turned out to have also some potential importance for the accuracy properties. When the matrix R, (8.6), is nearly rank-deficient the estimate $\hat{\theta}$ will have a bad accuracy (see Example 4.4).

The sufficient conditions derived in Chapter 4 are tied to different specific IV variants. In general terms it has then been assumed either that the input is white noise or that a certain filter depending on both the system and the IV variant is strictly positive real, see Theorems 4.2-4.4 for details. It should be stressed, though, that such conditions are sufficient but not necessary for consistency. In fact, the consistency condition (8.6) is under fairly weak conditions generically fulfilled (that is, most IV variants give consistency for *almost all* combinations of the system parameters), cf Theorem 4.1.

The accuracy properties of IV estimates were derived in Chapter 5. We have shown in Theorem 5.1 that under mild conditions the estimates are asymptotically gaussian distributed. To be more specific, assume that the consistency conditions (8.6), (8.7) are fulfilled. Then the (normalized) estimation errors $\sqrt{N}(\hat{\theta}-\theta^*)$ are asymptotically gaussian distributed with zero mean and covariance matrix given by

$$P_{IV} = (R^TQR)^{-1}R^TQE\{[\sum_{i=0}^{\infty} Z(t+i)K_i]\Lambda[\sum_{j=0}^{\infty} Z(t+j)K_j]^T\}QR(R^TQR)^{-1} \tag{8.13}$$

where R is given by (8.12) and $\{K_i\}_{i=0}^{\infty}$ are defined through

$$\sum_{i=0}^{\infty} K_i z^i \equiv F(z)H(z) \tag{8.14}$$

We have also shown, in Theorem 6.1, that the covariance matrix P_{IV} has an achievable lower bound

$$P_{IV} \geq P_{IV}^{opt} \triangleq \{E[H(q^{-1})^{-1}\tilde{\phi}^T(t)]^T\Lambda^{-1}[H(q^{-1})^{-1}\tilde{\phi}^T(t)]\}^{-1} \tag{8.15}$$

The consistency conditions (8.6), (8.7) and the accuracy result (8.13) make it possible, at least in principle, to evaluate any IV variant with respect to the consistency and accuracy properties of the corresponding estimate. The discussion on these and related aspects is deferred to the next section.

8.2 HINTS FOR CHOOSING THE IV VARIANT

The choice of the IV variant is indeed a central problem for the IV estimation, since both the consistency and the accuracy properties of the parameter estimates depend heavily on the IV scheme employed. The computational complexity is also determined by the IV variant used.

In general, the IV variant should be chosen by a trade-off between the corresponding properties of the parameter estimate and the associated computational burden. Below we discuss simple, as well as bootstrap and optimal IV variants. For each of these three types of IV variants of increasing complexity we mention briefly the properties that can be expected for the corresponding parameter estimates.

Simple IV variants

This type of IV variant is most commonly obtained by taking $F(q^{-1}) \equiv I$, and using delayed and possibly filtered input values to construct Z(t). In view of A7, the consistency condition (8.7) is then automatically fulfilled. The other consistency condition (8.6) is fulfilled *generically* under mild assumptions on the experimental conditions and the way the input enters into the IV matrix Z(t). For instance, the input signal must be persistently exciting of a sufficiently high order (cf. A6). Also, one should avoid using unnecessarily delayed inputs as instrumental variables.

The accuracy of the estimated parameters obtained with simple IVMs may, however, be poor in some situations. On the other hand, the computational requirements are quite moderate. The use of a simple IV variant may indeed lead to considerable saving of computation time. To illustrate this, let us assume that the matrix $\phi(t)$ is block diagonal

$$\phi^T(t) = \begin{bmatrix} \varphi_1^T(t) & & \bigcirc \\ & \ddots & \\ \bigcirc & & \varphi_{ny}^T(t) \end{bmatrix} \tag{8.16}$$

where $\{\varphi_i(t)\}$ are $n\theta_i$-vectors ($\sum_{i=1}^{ny} n\theta_i = n\theta$). Note that (8.16) holds for most parameterizations commonly used, e.g. the diagonal form of Example 2.4 and the full polynomial form of Example 2.5. Assume that also the IV matrix is constrained to have a diagonal structure, i.e.

$$Z(t) = \begin{bmatrix} z_1(t) & & \bigcirc \\ & \ddots & \\ \bigcirc & & z_{ny}(t) \end{bmatrix} \tag{8.17}$$

with $\{z_i(t)\}$ being nz_i-vectors ($nz_i \geq n\theta_i$, and $\sum_{i=1}^{ny} nz_i = nz$). Then the IV estimator (8.9) reduces to

$$\left[\sum_{t=1}^{N} z_i(t)\varphi_i^T(t)\right]\hat{\theta}_i = \sum_{t=1}^{N} z_i(t)y_i(t) \qquad i = 1,\dots, ny \tag{8.18}$$

with $\{\hat{\theta}_i\}$ being $n\theta_i$-vectors and $[\hat{\theta}_1^T \dots \hat{\theta}_{ny}^T] \triangleq \hat{\theta}^T$.

Instead of solving a large system of linear equations, i.e. (8.9), ny smaller systems are now to be solved. Since the number of arithmetic operations needed to solve a linear system is proportional to the cube of the number of unknowns, the saving in computer time may be substantial.

The computational burden can be further reduced if we use some special parameterizations for which $\varphi_i(t)$ does not vary with i, so that $\varphi_1(t) = \ldots = \varphi_{ny}(t) \triangleq \varphi(t)$. (This is true e.g. for the "full polynomial form" of Example 2.5). Similarly, take $z_1(t) = \ldots = z_{ny}(t) \triangleq z(t)$. The IV estimator (8.18) then reduces to

$$[\sum_{t=1}^{N} z(t)\varphi^T(t)][\hat{\theta}_1 \ldots \hat{\theta}_{ny}] = \sum_{t=1}^{N} z(t)y^T(t) \qquad (8.19)$$

Hence only one linear system of moderate dimension needs to be solved in this case. As compared to (8.9), the computational effort required by (8.19) is reduced about ny^3 times.

Note that the linear systems (8.18), (8.19) will in general be overdetermined. This was discussed to some extent in Examples 3.5, 3.6.

Bootstrap IV variants

The introduction of the bootstrap estimators could be seen simply as an attempt to get IV estimates that are consistent under fairly weak conditions. Recall that with a simple IV scheme only *generically* consistent estimates can in general be obtained. We showed in Example 3.2 that the following idealized IV variant

$$Z(t) = \tilde{\phi}(t,\theta^*) \qquad F(q^{-1}) \equiv I \qquad (8.20)$$

fulfils the consistency conditions under the general basic assumptions. The IV estimate corresponding to (8.20) cannot, of course, be implemented directly, since $\tilde{\phi}(t,\theta^*)$ contains the unmeasurable noise-free output (we indicated this by using θ^* as an argument for $\tilde{\phi}(\cdot)$). A simple idea to overcome this difficulty, is to use the following iterative algorithm (which we called BE_1 in Example 3.7)

$$\hat{\theta}^{k+1} = [\sum_{t=1}^{N} \tilde{\phi}(t,\hat{\theta}^k)\phi^T(t)]^{-1}[\sum_{t=1}^{N} \tilde{\phi}(t,\hat{\theta}^k)y(t)] \qquad (8.21)$$

Here $\hat{\theta}^k$ denotes the estimate obtained at iteration k. Since the basic idea behind the algorithm (8.21) is to swap between estimating the IV matrix and the parameter vector, (8.21) is often called a bootstrap estimator.

It was shown in Theorem 4.5 that the iterative algorithm (8.21) has the same nice consistency properties as the "idealized" IV estimator corresponding to (8.20). This essentially means that

$$\lim_{N\to\infty} \hat{\theta}^1 = \theta^*$$

(provided the initial estimate $\hat{\theta}^0$ is such that the inverse matrix in (8.21) exists). In particular one might then expect that (8.21) is globally convergent for sufficiently large N. Furthermore, the estimate given by (8.21) has the same asymptotic accuracy properties as the corresponding idealized estimate, see Theorem 5.1, 5.2 (and also the analysis in Section 6.4). This accuracy is not necessarily optimal in any sense, but we may expect that in general (8.21) will give a reasonable accuracy. This together with the improved consistency properties as compared to a simple IV estimator is the motivation for the additional computational effort required to use the bootstrap algorithm.

Finally, note that another bootstrap IV algorithm, which we called BE_2 in Example 3.7 was found (see Theorem 4.6) to have poor consistency properties. Thus it is less useful than the algorithm (8.21).

Optimal IV variants

We gave in (8.15) a lower bound for the covariance matrix of the parameter estimates. In Chapter 6 we saw that this lower bound can really be achieved by appropriately choosing the instruments Z(t) and the prefilter $F(q^{-1})$. Two different possibilities were given explicitly. One of these is the following

$$Z(t) = [\Lambda^{-1}H(q^{-1})^{-1}\tilde{\phi}^T(t)]^T$$

$$F(q^{-1}) = H(q^{-1})^{-1} \qquad (8.22)$$

$$\hat{\theta} = [\sum_{t=1}^{N} Z(t)\cdot H(q^{-1})^{-1}\phi^T(t)]^{-1}[\sum_{t=1}^{N} Z(t)\cdot H(q^{-1})^{-1}y(t)]$$

The other possibility is described in (6.20).

Now, (8.22) requires the undisturbed (unmeasurable) output x(t) as well as the noise autocorrelation (in form of Λ and $H(q^{-1})$) to be known. In practice an iterative technique must therefore be used to implement the optimal IV estimate. Note that the situation is more complex than in the previous subsection since now we must estimate also the noise properties. To implement (8.22) we therefore need to combine in an iterative manner the optimal IV method with a procedure for estimation of the noise parameters, as for instance in the following multistep algorithm of Section 6.3

Approximate implementation of optimal IV estimator

Step 1 Apply an "arbitrary" IV method in the model structure

$$y(t) = \phi^T(t)\theta+\varepsilon(t)$$

The resulting estimate will be denoted $\hat{\theta}_1$.

Step 2 Apply a prediction error method to estimate β (and Λ in the multivariable case) in the model structure

$$y(t) = \phi^T(t)\hat{\theta}_1+H(q^{-1},\hat{\theta}_1,\beta)\varepsilon(t,\hat{\theta}_1,\beta)$$

where $\hat{\theta}_1$ is given by Step 1, and where $H(q^{-1},\theta,\beta)$ with β being an additional parameter vector, is a model of the noise shaping filter $H(q^{-1})$. The resulting estimate will be denoted $\hat{\beta}_2$, $\hat{\Lambda}_2$.

Step 3 Compute the optimal IV estimate as given by (8.22) using $\hat{\theta}_1$ to form $\tilde{\phi}(t)$ and $\hat{\theta}_1$, $\hat{\beta}_2$ to form $H(q^{-1})$, and $\hat{\Lambda}_2$ substituting Λ. The resulting estimate will be denoted $\hat{\theta}_3$.

Step 4 If desired, Step 2 can be repeated using $\hat{\theta}_3$, instead of $\hat{\theta}_1$. The resulting estimate will be denoted $\hat{\beta}_4$, $\hat{\Lambda}_4$.

Some comments on the practical use and the properties of this algorithm are in order. The statements were verified in Section 6.4.

- In Step 1 we can use any simple IV method that gives a consistent estimate. For easy programming we may consider (8.22) with $H(q^{-1}) \equiv I$, $\Lambda = I$, and an initial estimate of θ^* containing only a few nonzero elements such that $Z(t)$ will be a full rank process.

- Several parameterizations of $H(q^{-1},\theta,\beta)$ are possible. The parameterization must be canonical in the sense that

$$H(z,\theta^*,\beta) \equiv H(z) \Rightarrow \beta = \beta^*$$

for a unique vector β^*. We call β^* the (vector of the) true noise parameters.

- The IV estimate $\hat{\theta}_3$ is asymptotically gaussian distributed with covariance matrix $P_{IV} = P_{IV}^{opt}$. It is thus (asymptotically) optimal within a fairly large class of IV methods.

- The maximal accuracy is obtained in Step 4. Moreover, further iterations will not change the asymptotic distribution of the parameter estimates. For a sufficiently large N, the multistep algorithm is therefore expected to converge in four steps only.

- In general the accuracy obtained with the multistep algorithm is inferior to the accuracy that can be achieved with a PEM. However, when the systems dynamics and the shaping filter of the additive disturbance w(t) are "decoupled" (that is to say, they can be modelled using independent parameter vectors), the multistep algorithm and the PEM give the same asymptotic accuracy. Furthermore, in such a case the multistep procedure converges in three steps only (at least for N large enough).

- As compared to a PEM, the multistep IV-based algorithm has some practical advantages. In the multistep procedure the estimation is partitioned into sub-problems of lower dimensions and hence the computational complexity will be smaller. Furthermore, the parameter estimates obtained with the multistep algorithm have nice uniqueness properties. The PEM may for some model structures give false estimates. The multistep algorithm could thus in some situations be viewed as an attractive alternative to PEM.

To summarize Section 8.2, we have discussed three classes of IVMs namely simple IVMs, BEs and optimal IVMs. These methods differ in their consistency properties, asymptotic accuracy and computational complexity.

The difference in consistency is not very essential. The only case when consistency may not occur is for basic IVMs. However, unless degenerated IVs are chosen, such as unnecessarily delayed inputs, consistency will occur generically.

The asymptotic accuracy is clearly best for optimal IVMs as compared to the other two classes. If accurate estimates are important the multistep algorithm for implementing the optimal IV estimate should be used.

Concerning the computational complexity, the simple IVMs have the smallest requirements. If fast calculations are the most important factor such an IVM, see e.g. Examples 3.1 and 3.5, should be used. For the optimal IVMs the noise parameters are estimated with an iterative algorithm. This can require a good deal of computation time. For some model structures, however, this part of the optimal IVM becomes a simple least squares fit. We will use such a model structure in Chapter 9. Then a reasonably fast algorithm is obtained.

8.3 HINTS FOR INPUT DESIGN

The input design problem is, to some extent, similar to that of choosing the IV variant. Indeed, similarly to the IV variant, the input signal used in an IV identification experiment, will influence both the consistency and the accuracy properties of the resulting parameter estimates. We could thus distinguish the following classes of input processes.

Inputs fulfilling the necessary condition for consistency

For linear systems, such inputs belong to the class of persistently exciting (p.e.) signals of a certain order. The actual order needed varies from case to case, depending on the system structure, the parameterization used etc. (see Examples 4.1, 4.2). If, for example, the input u(t) is a stationary process with rational non-singular spectral density (i.e. an ARMA process) then it is p.e. of any order. A pseudo-random binary sequence (PRBS), see Eykhoff (1974) for a general discussion, is another example of input signal often used in the identification of linear systems. It can be shown that a PRBS of period n and with the constant level substracted, is persistently exciting of order n-1, but not more, Ljung (1971).

For the nonlinear systems of Hammerstein type (see Examples 2.3 and 4.5) a necessary requirement on the input is that u(t) is a strongly persistently exciting signal of orders n and m [spe (n,m); see Appendix 1 for a definition]. Here, n depends on the order of the system, while m is determined by the degree of nonlinearity. Filtered white noise is spe (n,m) for any finite n and m, see Lemma A1.1.

Inputs fulfilling sufficient conditions for consistency

Here the possibilities seem to be more limited. It was, however, shown that under mild conditions a white input signal will guarantee the consistency of some simple IV estimates, for scalar and multivariable linear systems as well as for Hammerstein systems (see Theorems 4.2-4.3 and Example 4.5).

Optimal inputs

Inputs designed to maximize the accuracy of the parameter estimates were discussed in Chapter 7. To determine an optimal input we need, besides additional computer calculations, a priori information on the system. This will in general require a preliminary identification experiment which may be a quite expensive requirement. The use in practice of an optimal input must thus be clearly motivated by the expected increase of accuracy.

8.4 HINTS FOR MODEL STRUCTURE SELECTION

The choice of model structure is not an easy task. One reason is that this choice contains in itself several questions of various kinds to be answered. The purpose of this section is to illustrate how such questions can be attacked.

Let us first introduce the concepts *structural indices* and *classes of model structures*. With structural indices we mean integer parameters that describe the parameterization. A structural index can typically be a degree of a polynomial or a time delay. When varying the structural indices in a model structure we will say that a class of model structures is obtained.

Choice of class of model structures

How should the class of model structures be selected? The answer to this question will clearly depend on the final aim of the model and the available identification software. Also, if essential a priori information is available, then it should of course be taken into account when choosing the model class. However, such knowledge seldom exists in ways directly related to discrete time models. If no a priori information exists, then for single-output linear systems the parameter vector θ should consist of *all* the coefficients in the difference equation model as we illustrated in Examples 2.1-2.3. For multi-output systems the choice of model structure class is more complicated. We have discussed two specific possibilities in Chapter 2:

- A diagonal form was given in Example 2.4. Then $A(z,\theta)$ in (8.1) is a diagonal matrix. The structural indices are the $2 \cdot ny$ degrees $\{nai, nbi\}$ $i = 1,\ldots, ny$ (see Example 2.4).

- A full polynomial form was given in Example 2.5. Then all the elements of

$A_1(\theta),\ldots, A_{na}(\theta),\ldots, B_{nb}(\theta)$ see (8.2) are included in the parameter vector θ. The structural indices are na and nb.

For both these classes some computational advantages are at hand. The matrix $\phi(t)$, (8.3), is block diagonal and we can then arrange the computations so as to reduce the computer time needed (as described in Section 8.2).

The discussion so far concerns only the parameterization of the system dynamics. For the multistep algorithm the way the noise shaping filter $H(q^{-1},\theta,\beta)$ is parameterized deserves some comments as well. If a small algorithmic complexity is sought then $H(q^{-1},\theta,\beta)$ should be chosen so that Steps 2 and 4 become simple least squares problems. This means that for a SISO system the overall model structure will be chosen as follows (cf. (6.25))

$$A(q^{-1})y(t) = B(q^{-1})u(t)+ \frac{1}{D(q^{-1})} \varepsilon(t,\theta,\beta) \tag{8.23}$$

This class of model structure has been used for most of the case studies reported in Chapter 9.

A specific question when choosing the class of model structure is whether the system dynamics part and the noise part of the model should have some common parameters or not. The model (8.23) can e.g. be rewritten as,

$$y(t) = \frac{B(q^{-1})}{A(q^{-1})} u(t)+ \frac{1}{A(q^{-1})D(q^{-1})} \varepsilon(t,\theta,\beta) \tag{8.23'}$$

where it is clear that the two parts are constrained to have some common poles. Of course, any available a priori information should be used when choosing the class of model structures. If e.g. it is expected that common poles do exist (a reason for this can be that some disturbances enter early in the process) then the accuracy of the model will be improved if this is taken into account. On the other hand, a model structure without common parameters in the system dynamics part and the noise part can be expected to be more robust when applied to real systems, cf Ljung (1978), Ljung and Söderström (1983). Furthermore, for the multistep algorithm convergence is then for large data series expected to occur after 3 steps only.

To get general rules for the selection of the class of model structure is of course a very complex task especially when the system does not belong to the classes considered. Such a constraint must be faced in practice. We will not discuss this problem further. Let us only remark here that a preliminary study is given by Ljung and Söderström (1983). There it was examined how various low order model structures can approximate a high order system.

Choice of structural indices - theoretical aspects

Once the class of model structures has been selected we have to find appropriate values of the structural indices. This will be done in one way if we look at the problem from a theoretical point of view, and in a somewhat different way if practical aspects are crucial. In theory the problem is approached assuming that the general Assumptions A1-A7 can be fulfilled. As we have remarked before, in practice this is hardly the case since a real system is seldom linear and of finite order. From a practical point of view we thus have to handle the problem differently. Then it is more relevant to seek good approximations of the system rather than consistent estimates of the system parameters. We will defer a discussion of such practical aspects to the next subsection.

From a theoretical point of view it is necessary that the structural indices are determined so that the general assumptions are fulfilled. Assumptions A4, A5 are then the interesting ones. If the indices are chosen too small, then no θ^* will exist, since the system *S* will not belong to the model structure *M*. On the other hand if the indices are too large, θ^* may not be unique. Then the matrix R, (8.6), becomes singular and the linear system (8.9) giving the estimate will be illconditioned. Consistency can not be expected in that case. To exemplify, for a SISO model we obtained in Example 2.6 the following consistency condition

$$\min(na-na^*, nb-nb^*) = 0 \qquad (8.24)$$

where na^*, nb^* are the system orders. This relation describes clearly in what way the structural indices (here the model orders na, nb) are constrained to fulfil Assumptions A4, A5.

The other interesting point illustrated by (8.24) is that in general there will be many values of the structural indices that will fulfil A4, A5. Out of these values, those leading to the most "parsimonious" model should in general be chosen, as might be expected. Two reasons for this choice can be given.

- The computational complexity is then as small as possible both in the estimation and in the future use of the model since the minimal number of parameters is chosen. For multivariable systems it may, though, sometimes pay to estimate a few more parameters e.g. if in such a case the simple form (8.19) can be used.

- It could be expected from the "parsimony principle" that the best accuracy is obtained when as few parameters as possible are used. This is in fact true for prediction error methods, see Söderström et al (1974 b), and also for optimal IV methods. However, this is not always true for basic IV variants, see Stoica and Söderström (1982 e). The last point should not though be exaggerated. After all, the covariance matrix of the parameter estimates is proportional to 1/N. Thus, the parsimony principle should be considered as a good rule of thumb.

Supposing the general assumptions apply, the structural indices should in conclusion be taken as small as possible subject to the constraint that the true system belongs to the model structure.

We now continue to discuss means to find the smallest possible values of the structural indices.

In general terms, the AIC method of Akaike (1971, 1981) and its variants are amongst the most popular and most efficient ways to find the values of the indices. However, they rely heavily upon a PEM being used for the parameter estimation. For the application to IVMs we thus should seek for alternative ways. A natural procedure for model structure selection within the IV estimation, Wellstead (1978), Söderström and Stoica (1978), Stoica (1982 a), is based on the "instrumental product-moment matrix" R, see (8.12),

$$R = EZ(t)\cdot F(q^{-1})\tilde{\phi}^T(t,\theta^*) \tag{8.25}$$

R has the interesting property to be rank deficient when $S \in M$ but θ^* is not unique (i.e. the structural indices are chosen too high), cf Lemma 4.1. Furthermore, for even simple choices of the IV variant, R will be (at least generically) of full rank (equal to $n\theta$) when θ^* is unique (Assumption A5). The matrix R could thus be used to find appropriate values of the structural indices. In practice, however, R is not available and we should use a sample estimate, say $\hat{R}$, of R to make inferences on rank R. For some particular cases this is quite feasible, see, e.g. Stoica (1981 a) where scalar ARMA models are treated. In general, it may be too cumbersome to use $\hat{R}$ in a well-defined statistical way to test the null hypothesis rank $R < n\theta$. In such cases, the following alternative may be preferable, see Wellstead (1978), Young et al (1980). Choose $F(q^{-1}) \equiv I$ and $Z(t) = \tilde{\phi}(t,\hat{\theta})$, with $\hat{\theta}$ being an estimate of θ^*. Then vary the structural indices and evaluate det $(\hat{R})$, with

$$\hat{R} = \frac{1}{N}\sum_{t=1}^{N} \tilde{\phi}(t,\hat{\theta})\phi^T(t) \tag{8.26}$$

where $\tilde{\phi}(\cdot,\cdot)$ and $\phi(\cdot)$ correspond to a given set of structural indices. Alternatively, we could evaluate, for some suitable norm,

$$\text{cond } \hat{R} = \|\hat{R}\| \cdot \|\hat{R}^{-1}\|$$

The smallest structural indices leading to a "reasonably large" value of det $(\hat{R})$, ***and*** which are such that "small" values of det $(\hat{R})$ result if all indices are increased by at least one should be chosen. For cond $\hat{R}$, in the above phrase "large" should be replaced by "small" and viceversa.

Note that in (8.26) we could use some simple IV matrix instead of $\tilde{\phi}(t,\hat{\theta})$ thus avoiding the need for an estimate of θ^*. Practical experience shows, however, that the choice (8.26) will in general lead to better results as compared to other more or less ad hoc choices of the IV matrix, see Young et al (1980) for details. This seems indeed true for on-line applications of IVMs. For off-line applications, the use in (8.26) of a simple IV matrix instead of $\tilde{\phi}(t,\hat{\theta})$ may however be preferable, at least due to the saving in computer time that results.

Note also, that the above model structure selection procedure, while being quite simple, will apparently provide consistent estimates of the structural indices (that is to say, the probability to choose wrong indices with the procedure, tends to zero as N approaches infinity). For small or medium sample sizes, however, the procedure may often fail to select the "true" indices, Stoica (1981 a).

The structure testing procedure discussed above should be viewed in relation to subsequent IV parameter estimation exercises (although it can also be used independently of such exercises). On one hand, this means that the matrix $\hat{R}$ used for structure selection purposes should be precisely that appearing in the IV algorithm. On the other hand, note that application of the test procedure will essentially require the computation of $\hat{R}^{-1}$ for each of the tested structural indices so that with a moderate additional effort we can also get the corresponding IV models.

The IV models obtained for various structural indices may prove very useful for the model structure selection. Indeed, whenever M does not contain the system we naturally expect the corresponding model to have a poor performance. Further, when $S \in M$ but θ^* is not unique, (very) poor performance is again expected. In this last case, cond $\hat{R}$ is likely to be "large", and hence the linear system of equations giving the IV estimate will in general be ill-conditioned. A numerical routine computing the IV estimate may then give inaccurate (if not abnormal) values. By eliminating the model structures leading to (very) poor performances we may ascertain the "right" structure. This idea is developed in the following.

Choice of structural indices - practical aspects

For most processes encountered in practice our models are not exact descriptions but only approximations. Thus from a practical point of view the structural indices should apparently be chosen by a trade-off between the model complexity and its accuracy. Below we give two examples of how the accuracy of a model can be evaluated.

Example 8.1 Model accuracy functions

The "performance" of a model with parameters $\hat{\theta}$ can for instance be expressed using the following output error criterion

$$C_o = \{\frac{1}{N}\sum_{t=1}^{N} \| y(t)-y_M(t,\hat{\theta})\|^2\}^{1/2} \qquad (8.27)$$

where

$$y_M(t,\hat{\theta}) = A(q^{-1},\hat{\theta})^{-1}B(q^{-1},\hat{\theta})u(t)$$

is the model output. When also a noise model is estimated (as for the multistep algorithm) we can use the following prediction error criterion to assess the model performance

$$C_p = \{\frac{1}{N}\sum_{t=1}^{N} \| y(t)-\hat{y}_M(t|t-1,\hat{\theta})\|^2\}^{1/2} \qquad (8.28)$$

where the one step ahead prediction based on the model is given by

$$\hat{y}_M(t|t-1,\hat{\theta}) = [I-\hat{H}(q^{-1})^{-1}A(q^{-1},\hat{\theta})]y(t)+\hat{H}(q^{-1})^{-1}B(q^{-1},\hat{\theta})u(t) \qquad ■$$

The criteria C_o and C_p are expected to vary with the model structure as in Figure 8.1. This follows from the discussion in the previous subsection.

As compared to the test based on det $(\hat{R})$, or cond $\hat{R}$, the use of C_o, C_p, while requiring some additional computational effort, will presumably give clearer answers to the selection of structural indices. In particular, practical experience with simulated examples has shown that the model structures for which θ^* is not unique lead in general to <u>*very*</u> large values of C_o and C_p, or even to computer overflows. This may often be sufficient to find easily the "right" or "most appropriate" structure.

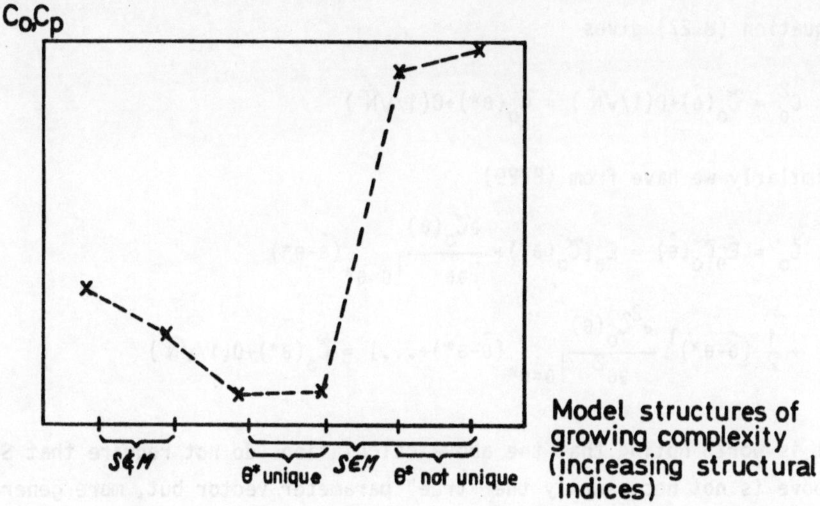

Figure 8.1. Typical variations of the accuracy functions C_o and C_p, with the model structure. (The relation $S \in M$ is here assumed to hold at least in an approximate sense).

Criteria like C_o, (8.27), and C_p (8.28) can also be used to approximately evaluate the accuracy of a *model structure*. When assessing the accuracy of a given model structure M it is natural to use criteria like

$$\bar{C}_o = E_{\hat{\theta}}[E\| y(t)-y_M(t,\hat{\theta})\|^2] \tag{8.29}$$

$$\bar{C}_p = E_{\hat{\theta}}[E\| y(t)-\hat{y}_M(t|t-1,\hat{\theta})\|^2] \tag{8.30}$$

In (8.29), (8.30) the expectation $E_{\hat{\theta}}$ is with respect to the distribution of $\hat{\theta}$. (Different values of $\hat{\theta}$ lead to different models in the given model structure).

Criteria like $\bar{C}_o$ and $\bar{C}_p$ cannot be used in practice since they cannot be evaluated without exact knowledge of the true system. Note, however, that C_o^2 (C_p^2) is a good approximation of $\bar{C}_o$ ($\bar{C}_p$). The approximation error is of magnitude $1/\sqrt{N}$. This can be seen as follows.

Define

$$\tilde{C}_o(\theta) = E\| y(t)-y_M(t,\theta)\|^2$$

Equation (8.27) gives

$$C_o^2 = \mathcal{C}_o(\hat{\theta})+O(1/\sqrt{N}) = \mathcal{C}_o(\theta^*)+O(1/\sqrt{N})$$

Similarly we have from (8.29)

$$\bar{C}_o = E_{\hat{\theta}}\mathcal{C}_o(\hat{\theta}) = E_{\hat{\theta}}[\mathcal{C}_o(\theta^*)+ \frac{\partial\mathcal{C}_o(\theta)}{\partial\theta}\Big|_{\theta=\theta^*}(\hat{\theta}-\theta^*)$$

$$+ \frac{1}{2}(\hat{\theta}-\theta^*)^T \frac{\partial^2\mathcal{C}_o(\theta)}{\partial\theta^2}\Big|_{\theta=\theta^*}(\hat{\theta}-\theta^*)+\ldots] = \mathcal{C}_o(\theta^*)+O(1/\sqrt{N})$$

It is worth noting that the above calculations do not require that $S \in M$. Thus θ^* above is not necessarily the "true" parameter vector but, more generally, it denotes $\lim_{N\to\infty} \hat{\theta}$. Similar calculations can be performed for C_p and $\bar{C}_p$.

Thus the criteria C_o and C_p seem to be reasonable means to select also the structural indices leading to the most accurate (w.r.t. $\bar{C}_o$ or $\bar{C}_p$) *model structure.*

Chapter 9

CASE STUDIES

9.1 GENERAL ASPECTS

In this chapter we will present four comprehensive IV modeling applications using real data. We mentioned in connection with Table 3.1 that many applications of IV estimation techniques have been made to real data and processes. We specify some of these in Table 9.2 below.

We now give some general comments on the applications treated in this chapter. We first present a short summary of the IV variants and the model structures used for each case study. This information is contained in Table 9.1

Case study	System	N	Model structure	IV-variant
Drum drier (Section 9.2)	SISO	200	$Ay = Bu + \frac{1}{D}\varepsilon$	Multistep algorithm
Economic process (Section 9.3)	SISO	120	$Ay = Bu + \frac{1}{D}\varepsilon$	Multistep algorithm
Gas furnace (Section 9.4)	SISO	296	$Ay = Bu + \frac{1}{D}\varepsilon$	Multistep algorithm
Turbo-alternator (Section 9.5)	MIMO	100	Diagonal form $A_i y_i = B_{1i} u_1 + B_{2i} u_2 + \varepsilon$, $i = 1,2$	z with delayed inputs as elements

Table 9.1. Overview of the case studies.

Process	
1. <u>Technical/industrial processes</u>	
aircraft engine	Wellstead and Rojas (1982)
cement kiln	Stoica and Söderström (1981 d)
distillation column	Gauthier and Landau (1978) Young et al (1971)
drum drier	this book (Section 9.2)
gas furnace	Young and Jakeman (1979 b) Young et al (1980) Young et al (1971) this book (Section 9.4)
heat exchanger	Bauer and Unbehauen (1978)
missile	Young et al (1971)
motor-alternator	Chan (1973)
paper machine	Gentil et al (1973)
steam superheater	de la Puente and Albertos (1979)
turbo-alternator	Sinha and Caines (1977) this book (Section 9.5)
2. <u>Nontechnical processes</u>	
BOD-DO (water quality)	Whitehead and Young (1979) Young and Whitehead (1977)
economic processes	Young and Jakeman (1979 b) Young et al (1971) this book (Section 9.3)
fluorescence decay	Young and Jakeman (1979 b)
human circulatory system	Young et al (1980)
ozone level	Jakeman et al (1980) Young and Jakeman (1979 b)
rainfall-flow	Whitehead et al (1979) Young (1974)

Table 9.2 Survey of some references reporting real data applications of IVMs.

For the SISO systems we thus use the model structure

$$A(q^{-1})y(t) = B(q^{-1})u(t-n\tau) + \frac{1}{D(q^{-1})}\varepsilon(t) \tag{9.1}$$

where $n\tau$ denotes the dead time, and where

$$\begin{aligned} A(q^{-1}) &= 1+a_1q^{-1}+\ldots+a_{na}q^{-na} \\ B(q^{-1}) &= \quad b_1q^{-1}+\ldots+b_{nb}q^{-nb} \\ D(q^{-1}) &= 1+d_1q^{-1}+\ldots+d_{nd}q^{-nd} \end{aligned} \tag{9.2}$$

The (unknown) parameters describing the model set (9.1) are collected in the following two vectors

$$\begin{aligned} \theta &= [a_1 \ldots a_{na} \; b_1 \ldots b_{nb}]^T \\ \beta &= [d_1 \ldots d_{nd}]^T \end{aligned} \tag{9.3}$$

For given na, nb, nd and $n\tau$, the parameter vectors θ and β will be estimated from the data using the multistep algorithm. This IV-based algorithm was described in detail in Chapter 6.

Note that for the model structure (9.1), Step 2 (and similarly Step 4) becomes very simple since it is just a simple least squares fit. It can be described by

$$\hat{\beta} = [\sum_{t=1}^{N} \psi(t)\psi^T(t)]^{-1}[\sum_{t=1}^{N} \psi(t)\hat{v}(t)] \tag{9.4}$$

where

$$\psi(t) = [-\hat{v}(t-1)\ldots -\hat{v}(t-nd)]^T$$

$$\hat{v}(t) = \hat{A}(q^{-1})y(t)-\hat{B}(q^{-1})u(t-n\tau)$$

The simple form (9.4) of Steps 2 and 4 was in fact the main reason for choosing the model structure (9.1) in the SISO case studies.

We proved in Chapter 6 that when the system belongs to the model structure and the number of data is "large", the multistep algorithm is expected to converge after four steps only. To show that this theoretical result may well apply in practice, where none of the above assumptions is exactly valid, we shall present the estimates obtained with the multistep algorithm after 4 and 10 steps. The 4-step algorithm will be labelled 4SA and, similarly, we shall use the abbreviation 10SA to designate the 10-step algorithm (cf. also Section 6.4).

The obtained models will be compared using the criteria C_0, (8.27), and (when applicable) C_p, (8.28). Such comparisons are being used to select the final model structure, i.e. for (9.1) the integers na, nb, nd and nτ. For the final models selected, we will also present plots of the input-output signals together with the model outputs.

The real data series contain nonzero mean values. In all equations as well as in the estimation, the arithmetic means have been subtracted from the involved signals. In the plots, however, the nonzero mean values are kept.

9.2 A DRUM DRIER

The process to be identified is a control loop of a drum drier for green forage described by Modén (1981), Modén and Nybrant (1980). The process is depicted in Figure 9.1.

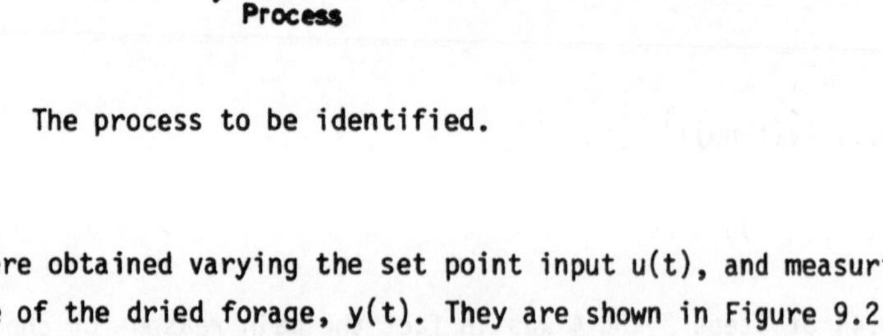

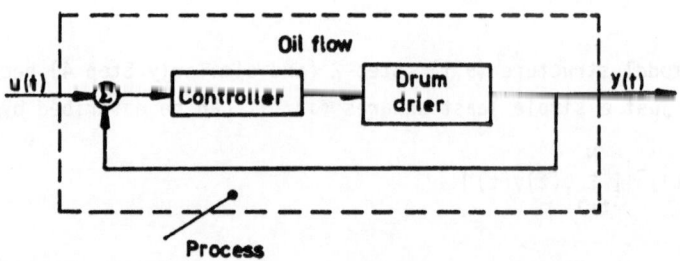

Figure 9.1. The process to be identified.

The data were obtained varying the set point input u(t), and measuring the resulting temperature of the dried forage, y(t). They are shown in Figure 9.2. The input signal u(t) consists of a manually adjusted value to which a PRBS was added.

The 4SA and 10SA were applied in a number of model structures within the general class (9.1). In Table 9.3 we have given the values of C_p and C_0 for a number of cases.

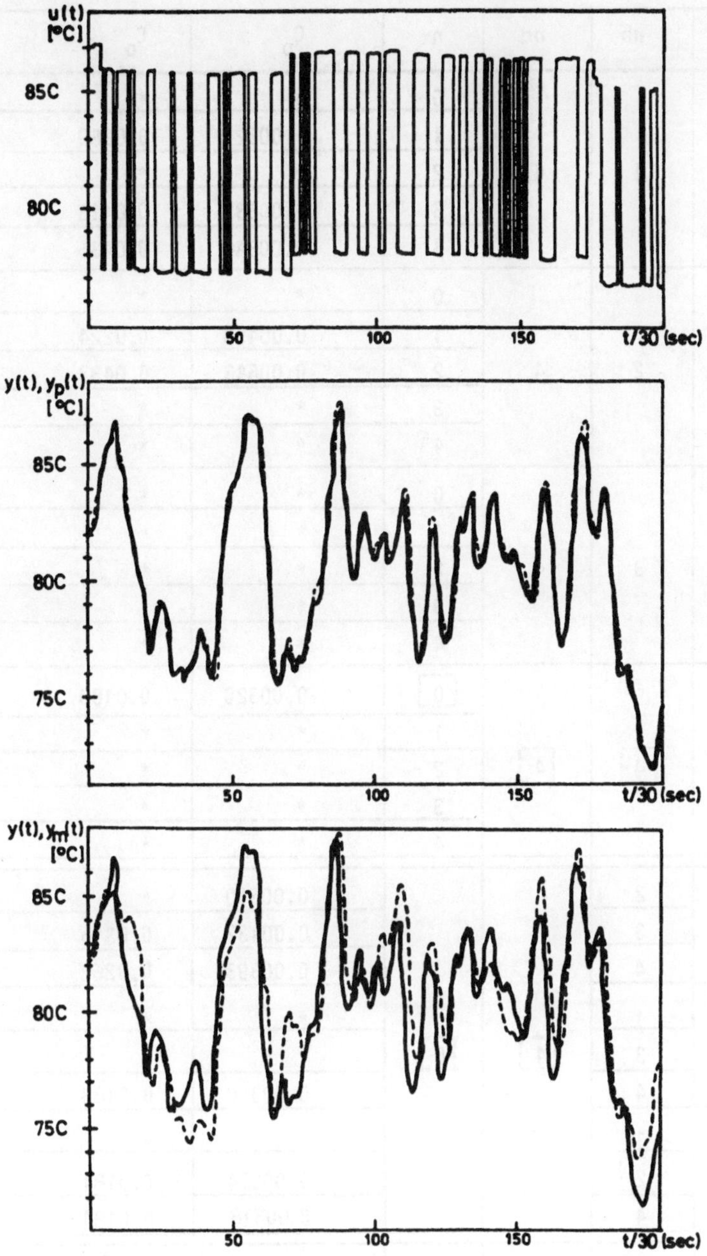

Figure 9.2. Plots of the signals for the forage drier. The model is given by (9.5). The signals shown are i) the input u(t); ii) the measured output y(t) and the predicted output $y_p(t)$; iii) y(t) and the model output $y_m(t)$. The mean values are $\bar{u} = 81.8^oC$, $\bar{y} = 80.4^oC$.

na	nb	nd	nτ	C_p	C_o
1	1	4	0	*	*
			1	0.00521	0.0389
			2	*	*
			3	0.00639	0.0418
			4	0.00846	0.0436
2	2	4	0	*	*
			1	0.00450	0.0224
			2	0.00643	0.0433
			3	*	*
			4	*	*
3	3	4	0	*	*
			1	*	*
			2	*	*
			3	*	*
			4	*	*
[4]	[4]	[4]	[0]	0.00320	0.0180
			1	*	*
			2	*	*
			3	*	*
			4	*	*
1	2	[4]	[0]	0.00510	*
	3			0.00437	0.0316
	4			0.00393	0.0262
2	1			*	*
	3			*	*
	4			0.00318	0.0183
[3]	1			*	*
	[2]			0.00324	0.0181
	4			0.00318	0.0184
[3]	[2]	3	[0]	0.00324	0.0181
		2		0.00333	0.0181
		1		0.00337	0.0183
		[0]		0.00340	0.0181

Table 9.3. Drum drier application: The model structures tested and the associated criteria. An asterisk indicates a large value of the criterion (typically at least 100 times larger than the minimal value).

It follows from the table that the following values of the structure parameters are appropriate to use: $\hat{na} = 3$, $\hat{nb} = 2$, $\hat{nd} = 0$ and $\hat{n\tau} = 0$. Some comments on this choice are in order. First, in the second part of Table 9.3, the structure na = 2, nb = 4, nd = 4, $n\tau$ = 0 appears as a good alternative to that chosen. The structure na = 3, nb = 2, nd = 4, $n\tau$ = 0 was however preferred since it has one parameter less. Second, from a theoretical point of view it is somewhat striking that while the structure na = 3, nb = 2, nd = 4, $n\tau$ = 0, led to a good model (see part two of Table 9.3), the structure na = 3 = nb, nd = 4, $n\tau$ = 0 led to large values of C_o, C_p (cf. the first part of Table 9.3). This may, however, be explained in the following way. The (discrete-time) model of the process under investigation has one pole close to the unit circle (see below). Then it may be possible that the increase of nb from 2 to 3 led to fluctuations in the estimated parameters that made the model unstable.

The model (corresponding to the structure chosen) obtained with 4SA it the following

$$(1-1.475q^{-1}+0.461q^{-2}+0.114q^{-3})y(t) = (0.063q^{-1}+0.096q^{-2})u(t)+\varepsilon(t) \qquad (9.5)$$

The 10SA gave a similar result, namely

$$(1-1.474q^{-1}+0.444q^{-2}+0.133q^{-3})y(t) = (0.064q^{-1}+0.096q^{-2})u(t)+\varepsilon(t) \qquad (9.6)$$

Figure 9.2 shows the output $y_m(t)$ of the model (9.5), as well as its one-step ahead predicted output, $y_p(t)$. We use here $y_m(t)$ and $y_p(t)$ instead of the more detailed notations previously used $y_M(t,\hat{\theta})$ and $\hat{y}_M(t|t-1,\hat{\theta})$, respectively. The measured output y(t) is also plotted in Figure 9.2. It can be seen that the model gives a good fit to the data. The prediction error $y(t)-y_p(t)$ is quite small, although the (estimated) disturbance $y(t)-y_m(t)$ is substantial.

Interestingly enough, the equation errors of (9.5) form a white noise sequence. This suggests that the process can be equally well identified using the LSM. Applying the LS procedure - in various model structures - to the data, we have arrived at the following model

$$(1-1.512q^{-1}+0.497q^{-2}+0.110q^{-3})y(t) = (0.064q^{-1}+0.092q^{-2})u(t)+\varepsilon(t) \qquad (9.7)$$

which is similar to the IV model (9.5).

9.3 AN ECONOMIC PROCESS

This application is based on data taken from Makridakis and Wheelwright (1978). Figure 9.3 shows the variation of the weekly production and billings of a company. The company is interested in forecasting the value of its billings and since the production is a "leading indicator" for billings, a dynamic model relating billings (as output) to production (as input) should be determined. We will assume the structure (9.1) for this model. The multistep estimation procedure was applied for the various combinations of orders and dead time shown in Table 9.4. For the models so obtained the criteria C_p and C_o were computed.

From the first part of Table 9.4 it follows that we should retain the combination na = nb = 2, nd = 4 and $n\tau$ = 2. The second part shows that the values of na and nb should not be smaller than 2. Finally, it follows from the last part that there are no sensible differences between the models with na = nb = $n\tau$ = 2 and nd varying from 1 to 4. Then we take $\hat{nd}$ = 1 and have thus arrived at the following model structure: $\hat{na}$ = 2, $\hat{nb}$ = 2, $\hat{nd}$ = 1 and $\hat{n\tau}$ = 2.

The corresponding model obtained with the 4SA is

$$(1-1.531q^{-1}+0.755q^{-2})y(t) = (2.119q^{-3}-1.950q^{-4})u(t) + \frac{1}{1+0.446q^{-1}}\,\varepsilon(t) \qquad (9.8)$$

The model with the same structure obtained with the 10SA is

$$(1-1.504q^{-1}+0.722q^{-2})y(t) = (2.073q^{-3}-1.914q^{-4})u(t) + \frac{1}{1+0.404q^{-1}}\,\varepsilon(t) \qquad (9.9)$$

There are no essential differences between the parameters of (9.8) and (9.9). Furthermore, the two above models can be shown to be close to the model of Makridakis and Wheelwright (1978), obtained using the PEM. That model is given by

$$y(t) = \frac{2.105q^{-3}-1.976q^{-4}+0.05q^{-5}}{1-1.545q^{-1}+0.790q^{-2}}\,u(t) + \frac{1}{1-0.963q^{-1}+0.163q^{-2}+0.326q^{-3}}\,\varepsilon(t) \qquad (9.10)$$

While the structure of (9.10) on one hand, and that of (9.8) or (9.9) on the other, are different, straightforward calculations show that (9.8) can be rewritten as

$$y(t) = \frac{2.119q^{-3}-1.950q^{-4}+0.00q^{-5}}{1-1.531q^{-1}+0.755q^{-2}}\,u(t) + \frac{1}{1-1.085q^{-1}+0.072q^{-2}+0.337q^{-3}}\,\varepsilon(t) \qquad (9.11)$$

which is similar to (9.10). Note that the PE model (9.10) of Makridakis and

na	nb	nd	nτ	C_p	C_o
1	1	4	0	*	*
			1	*	*
			2	2.286	6.151
			3	3.689	6.505
			4	*	*
2	2	4	0	*	*
			1	2.116	5.514
			2	1.775	3.108
			3	3.721	7.460
			4	*	*
3	3	4	0	2.050	5.147
			1	1.754	3.083
			2	*	*
			3	*	*
			4	*	*
4	4	4	0	1.735	3.047
			1	*	*
			2	*	*
			3	*	*
			4	*	*
1	2	4	2	2.374	5.506
2	1			2.224	5.480
2	2	0	2	2.144	3.289
		1		1.893	3.282
		2		1.854	3.101
		3		1.815	3.226

Table 9.4. The model structures tested for the economic process, and the corresponding criteria. (* denotes a very large value of the criterion).

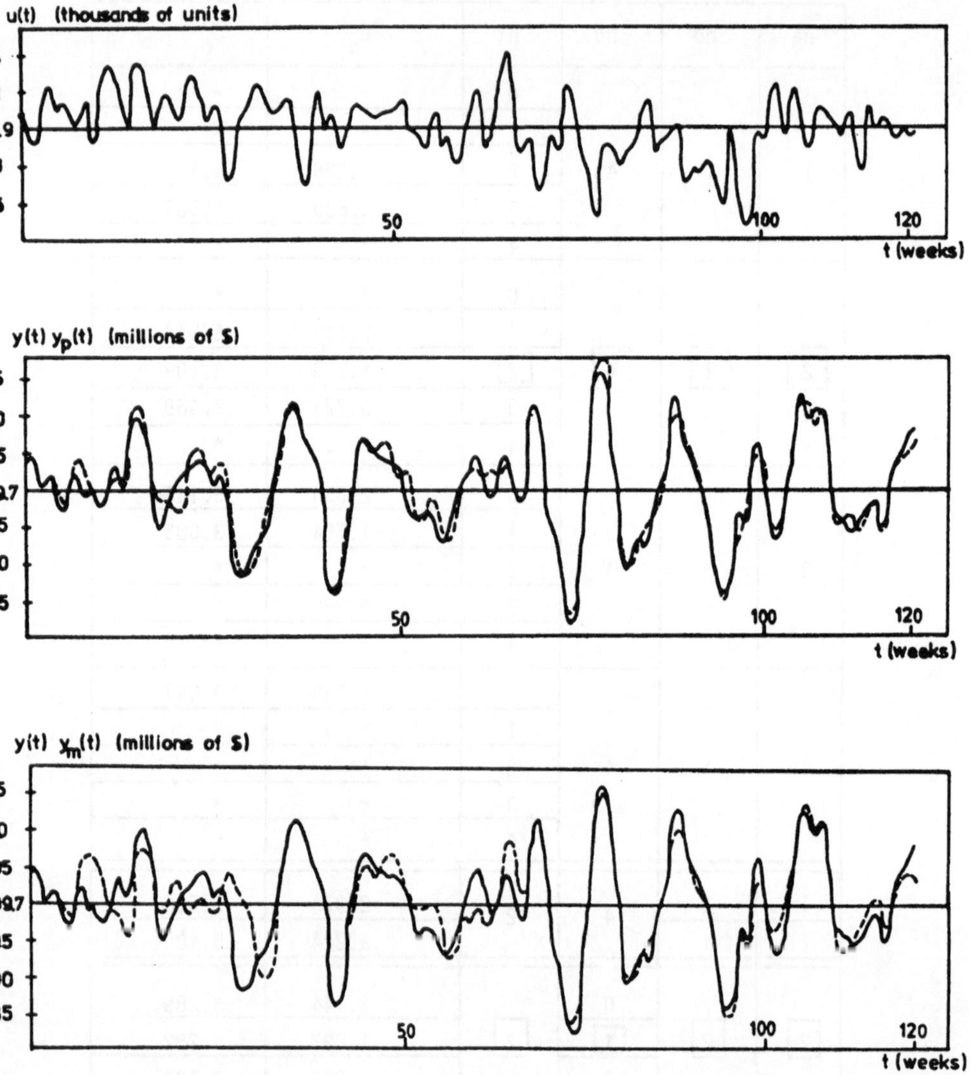

Figure 9.3. Plots of the signals for the economic process. The model used is given by (9.8). The signals shown are i) the input u(t); ii) the measured output y(t) and the predicted output $y_p(t)$; iii) y(t) and the model output $y_m(t)$. The mean values are $\bar{y} = 99.7$, $\bar{u} = 49.88$.

Wheelwright (1978) is clearly more complex than (9.8) or (9.9). Its structure does not account for common parameters between the system and noise transfer functions. It seems that for the process under study it is indeed appropriate to use common parameters. Failing to recognize this will make the model unnecessarily complicated.

Finally, the output of the model (9.8) as well as its one-step predicted output are shown in Figure 9.3 together with the process output. It can be seen that the model obtained with the multistep procedure gives quite good a fit to the data.

9.4 A GAS FURNACE

The basis for this application are the data given as series J in Box and Jenkins (1976). These data were continuously collected from a gas furnace and then read at every 9 seconds. The air feed of the furnace was kept constant, but the methane feed rate was varied and the resulting CO_2 concentration in the off gases was measured. The data so obtained are shown in Figure 9.4.

Using a PEM, Box and Jenkins obtained the following model for the process

$$y(t) = -\frac{0.53q^{-3}+0.37q^{-4}+0.51q^{-5}}{1-0.57q^{-1}}u(t) + \frac{1}{1-1.53q^{-1}+0.63q^{-2}}\varepsilon(t) \tag{9.12}$$

The criteria C_p and C_o associated to (9.12) have the following values

$$C_p = 0.238 \qquad C_o = 0.832 \tag{9.13}$$

In order to identify the gas furnace we have applied the multistep algorithm, in various model structures of the form (9.1), to the data. For na = 1, nb = 3, nd = 2 and $n\tau$ = 2 we have obtained the following model. (The 4SA and the 10SA give here results that coincide with at least two digits).

$$(1-0.54q^{-1})y(t) = -(0.52q^{-3}+0.38q^{-4}+0.52q^{-5})u(t) + \frac{1}{1-1.03q^{-1}+0.199q^{-2}}\varepsilon(t) \tag{9.14}$$

which can easily be rewritten as

$$y(t) = -\frac{0.52q^{-3}+0.38q^{-4}+0.52q^{-5}}{1-0.54q^{-1}}u(t) + \frac{1}{1-1.57q^{-1}+0.755q^{-2}+0.107q^{-3}}\varepsilon(t) \tag{9.15}$$

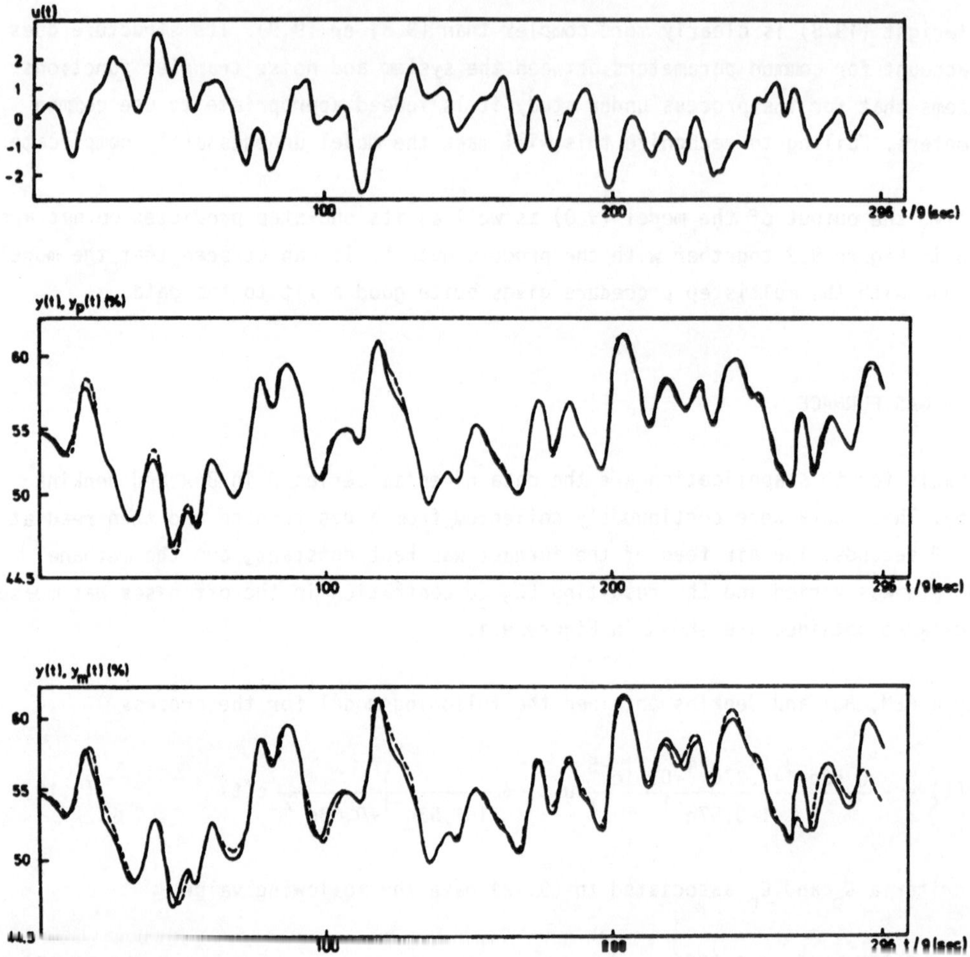

Figure 9.4. Plots of the signals for the gas furnace process. The model (9.17) was used. The signals shown are i) the input $u(t)$; ii) the measured output $y(t)$ and the predicted output $y_p(t)$; iii) $y(t)$ and the model output $y_m(t)$. The mean values are $\bar{y} = 53.5$, $\bar{u} = -0.65$.

Note that (9.12) and (9.15) are similar, even though they were obtained with different estimation methods applied in different model structures. The criteria C_p and C_o have for (9.14) the values

$$C_p = 0.238 \qquad C_o = 0.833 \tag{9.16}$$

which are very close to (9.13).

However, there are simpler model structures, which give models with similar performances. For instance, the 4SA applied in the structure $\hat{na} = 1$, $\hat{nb} = 2$, $\hat{nd} = 1$ and $\hat{n\tau} = 2$ gave the following model

$$(1-0.66q^{-1})y(t) = -(0.435q^{-3}+0.678q^{-4})u(t) + \frac{1}{1-0.784q^{-1}}\varepsilon(t) \tag{9.17}$$

for which

$$C_p = 0.249 \qquad C_o = 0.833 \tag{9.18}$$

While the values (9.16) and (9.18) are close to each other, the model (9.17) has two parameters less than (9.14) and should thus be preferred. For the same structure, the 10SA gave a very similar model, namely

$$(1-0.66q^{-1})y(t) = -(0.434q^{-3}+0.68q^{-4})u(t) + \frac{1}{1-0.785q^{-1}}\varepsilon(t) \tag{9.19}$$

The output of the model (9.17) as well as its one-step predicted output are compared with the process output in Figure 9.4. The approximation seems to be quite good.

Now, if (9.17) would be the "right" structure, then models with $na > 1$, $nb > 2$ ($n\tau = 2$) should normally lead to large values of C_p and C_o (or even to computer overflows), as was explained in Section 8.4 (cf also Table 9.4). For the gas furnace data, however, the criteria C_o and C_p were not substantially larger than (9.18) when higher order models were tried. This suggests that the gas furnace data do not fulfil all the assumptions used in the theory. It is, for instance, likely that the gas furnace, as many other processes, has dynamic characteristics which are time-varying.

We will show that there are good reasons to describe the gas furnace with a time-varying model. The multistep procedure was applied to the following sets of data (labelled by their positions within the total set of 296 data): 1 ÷ 70, 71 ÷ 140, 141 ÷ 210 and 211 ÷ 296. Each model was assumed to have the structure of (9.17).

The results obtained are shown in Table 9.5. By linearly interpolating the estimated values given in Table 9.5 it is shown in Figure 9.5 how the process parameters vary with time.

Parameter	Estimated values							
	Data set: 1÷70		Data set: 71÷140		Data set: 141÷210		Data set: 211÷296	
	4SA	10SA	4SA	10SA	4SA	10SA	4SA	10SA
a_1	-0.625	-0.627	-0.657	-0.658	-0.655	-0.655	-0.636	-0.627
b_1	-0.914	-0.906	-0.779	-0.779	-0.453	-0.452	0.749	0.767
b_2	-0.276	-0.276	-0.327	-0.325	-0.753	-0.754	-1.841	-1.854
d_1	-0.512	-0.516	-0.313	-0.313	-0.549	-0.549	-0.901	-0.903
$\bar{u}$	0.800		-0.333		-0.155		-0.419	
$\bar{y}$	50.77		54.27		53.46		55.15	

Table 9.5. Comparison of the models fitted to each of the four parts of the gas furnace data string. $\bar{u}$ and $\bar{y}$ denote the arithmetic means of $u(t)$ and $y(t)$ respectively.

Some interesting facts should be noticed:

- It can be seen from the table that the models obtained with 4SA and 10SA are quite similar despite the short length of the data series (N = 70).

- The process is likely to have time-varying dynamics. The large variations of the $\{\hat{b}_i\}$-parameters cannot be explained by sample fluctuations (which for the present case should be smaller than 0.3, see Box and Jenkins (1976)). Also, by applying the same technique as above to the following data sets: 1 ÷ 100, 101 ÷ 200 and 201 ÷ 296, we have obtained very similar results, see Figure 9.5. This gives further support to the view that the variations of the $\{\hat{b}_i\}$-estimates are due to a real change of the process parameters and not to sample fluctuations.

- The pole of the process is constant. The noise model parameter is also rather time-invariant. Instead, the $\{\hat{b}_i\}$- parameters change considerably. The model is minimum-phase for the first half of the record and becomes non-minimum phase afterwards. Despite these drastic changes in the $\{\hat{b}_i\}$-parameters, the gain of the model remains almost constant over the entire period of time. Without

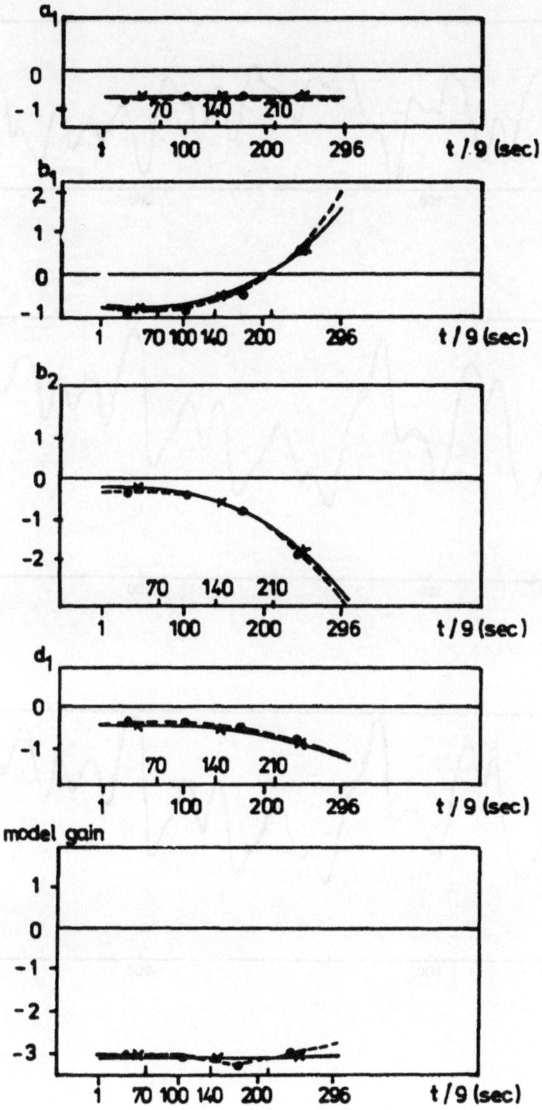

Figure 9.5. Time variation of the gas furnace model parameters. The curves are obtained through linear interpolation of the estimated values:

- • four sets of about 70 data pairs each
- x three sets of about 100 data pairs each

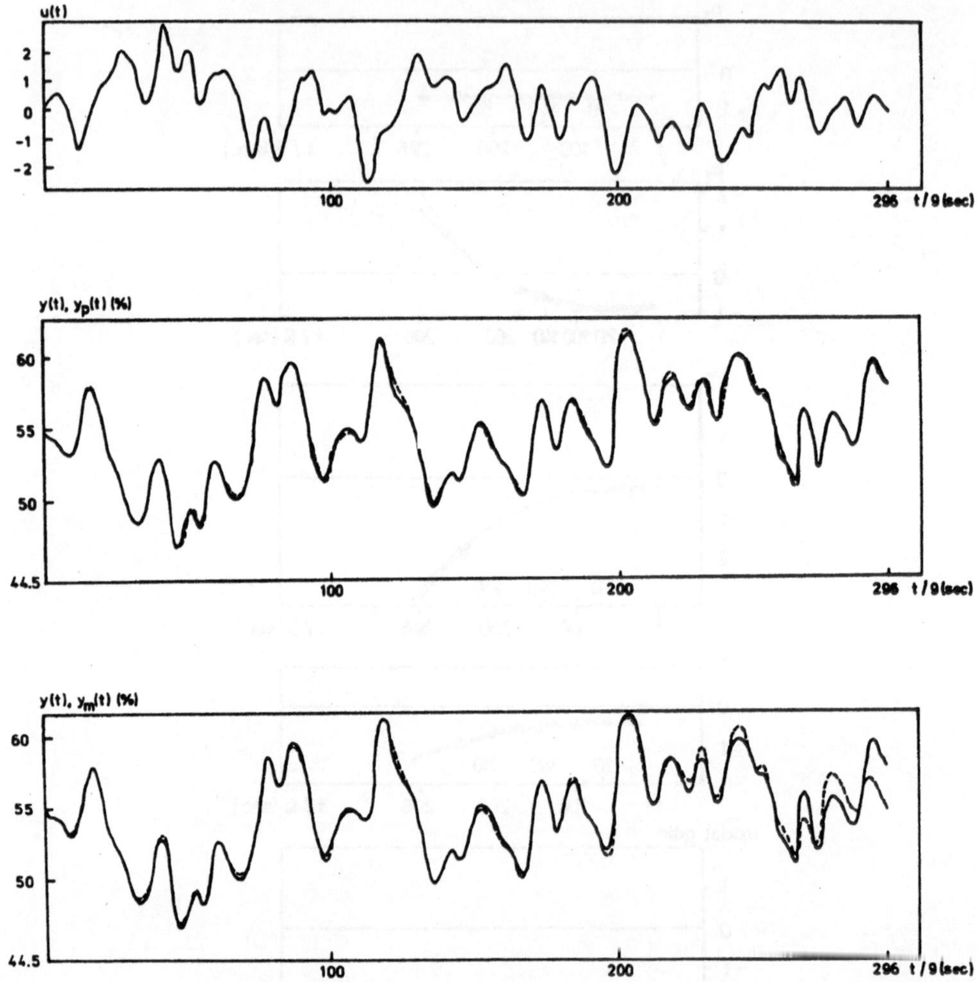

Figure 9.6. Plots of the signals for the gas furnace process. The time variable model given in Table 9.5 was used. The signals shown are i) the input u(t); ii) the measured output y(t) and the predicted output $y_p(t)$; iii) y(t) and the model output $y_m(t)$. The methane gas feed (in cubic feet per minute) is 0.6-0.04u(t). The mean values are $\bar{u} = -0.65$, $\bar{y} = 53.5$.

physical insight into the process it is difficult to give any definite explanation for this behaviour. However, a possible reason may be that the process has a time-varying dead time. That would lead, at least qualitatively, to the observed result.

In order to show that the model with parameters modified for different time periods accordingly to Table 9.5 can explain the data better than (9.17), the model output, the predicted output and also the associated criteria C_o and C_p were computed. The output of the model and the one step prediction are shown in Figure 9.6. The values of the criteria are

$$C_p = 0.201 \qquad C_o = 0.725 \tag{9.20}$$

They are really smaller than (9.18). Clearly, both (9.20) and Figure 9.6 support the hypothesis that the process can be better described with a time-varying model.

9.5 A TURBO-ALTERNATOR

The process is a 50 megawatt turbo-alternator described in Jenkins and Watts (1969), where also the results of a frequency-domain identification experiment are presented. Jenkins and Watts used a set of 1000 data points obtained at a sampling rate of 0.5 cps. However, only the first 100 data points which are explicitly given in Jenkins and Watts (1969) will be used here. We may remark that the same data set was used by Sinha and Caines (1977) to test IV estimation methods based on the direct parameterization of the model transfer function (that is to say, each entry of the transfer function matrix is parameterized independently of the other ones). Such an approach may give quite poor estimates, as illustrated by Sinha and Caines (1977), cf also a discussion in Stoica and Söderström (1981 d).

The turbo-alternator process can be described as a system with two inputs and two outputs, namely

u_1 the in-phase current

u_2 the out-of-phase current

y_1 the voltage amplitude

y_2 the voltage frequency

Figure 9.7 shows the input-output measurements. The signals depicted are (filtered) normal operating deviations from nominal values.

The model that will be fitted to the data is the vector difference equation (8.1) with the matrix polynomial $A(q^{-1},\theta)$ in diagonal form. The model for the first output $y_1(t)$ will therefore have the following structure

$$A(q^{-1})y_1(t) = B_1(q^{-1})u_1(t-n\tau_1)+B_2(q^{-1})u_2(t-n\tau_2)+\varepsilon(t) \tag{9.21}$$

The model for $y_2(t)$ will be similarly defined. In (9.21) $n\tau_i$ denotes the additional time delay for the i:th input, and

$$A(q^{-1}) = 1+a_1q^{-1}+\ldots+a_{na}q^{-na}$$

$$B_i(q^{-1}) = b_{i1}q^{-1}+b_{inb}q^{-nb} \qquad i = 1,2$$

For the estimation of the unknown parameters of (9.21) we shall use a simple IV method described in Section 4.3, the essence of which is repeated below for easy reference:

$$\theta = [a_1 \ldots a_{na}\ b_{11}\ b_{21} \ldots b_{1nb}\ b_{2nb}]^T \tag{9.22}$$

$$[\sum_{t=1}^{N} z(t)\varphi^T(t)]\hat{\theta} = [\sum_{t=1}^{N} z(t)y_1(t)] \tag{9.23}$$

$$z(t) = [u^T(t-1)\ldots u^T(t-na-nb)]^T$$

$$\varphi(t) = [-y_1(t-1)\ldots -y_1(t-na)\ u^T(t-1)\ldots u^T(t-nb)]^T$$

The method of orthogonal transformations, see e.g. Stewart (1973), was used to solve the overdetermined linear system (9.23).

Since the model structure is not a priori known, we solved (9.23) for various combinations of orders {na,nb} and dead times $\{n\tau_1,n\tau_2\}$. For each model so obtained we computed the criterion C_o, here denoted C_{o1} for output y_1 and C_{o2} for output y_2. The "best" model structure will then be chosen by a trade-off between the model complexity and its accuracy as measured by C_{o1}, C_{o2}.

For both outputs, y_1 and y_2 we have tested various model structures with

$$na \le 6 \quad nb \le 6 \quad n\tau_1 \le 3 \quad n\tau_2 \le 3.$$

For the first output, only the model structures shown in Table 9.6 were found to give comparative results. The other model structures gave (much) larger values of the loss function.

na	nb	$n\tau_1$	$n\tau_2$	C_{o1}
1	5	0	0	0.085
		0	2	0.095
1	6	0	0	0.064
		0	2	0.082
2	1	0	0	0.096
2	3	0	2	0.095
[2]	[5]	[0]	[0]	0.067

Table 9.6. Variation of the output error criterion C_{o1} with the model structure.

It can be seen from the table that the structure with na = 1, nb = 6, $n\tau_1 = n\tau_2 = 0$ and that with $\hat{na} = 2$, $\hat{nb} = 5$, $\hat{n\tau}_1 = \hat{n\tau}_2 = 0$ gave close results. The last structure contains, however, one parameter less and may be preferred. The corresponding model is given by

$$(1-0.95q^{-1}+0.299q^{-2})y_1(t) = (-0.40q^{-1}+0.503q^{-2}-0.446q^{-3}+0.109q^{-4}+0.047q^{-5})u_1(t)$$

$$+ (-0.451q^{-1}+0.827q^{-2}-0.809q^{-3}+0.405q^{-4}-0.057q^{-5})u_2(t)+\varepsilon(t) \qquad (9.24)$$

The output of the model (9.24) is compared to the real process output in Figure 9.7. The model gives a relatively good fit to the data.

Consider next the second output $y_2(t)$. Here a quite large number of model structures give comparative values of the loss function, see Table 9.7, and choosing the "best" structure is by no means an easy operation. However, noticing that a decrease in the loss function of, say, 0.01 is not negligible [taking into account the small fluctuations of $y_2(t)$], the structure

$$\hat{na} = \hat{nb} = 3 \qquad \hat{n\tau}_1 = \hat{n\tau}_2 = 1 \qquad (9.25)$$

may possibly be preferred. The corresponding model

$$(1-1.061q^{-1}+0.042q^{-2}+0.386q^{-3})y_2(t) = (0.214q^{-2}-0.668q^{-3}+0.354q^{-4})u_1(t)$$

$$+ (-0.032q^{-2}+0.090q^{-3}-0.031q^{-4})u_2(t)+\varepsilon(t) \qquad (9.26)$$

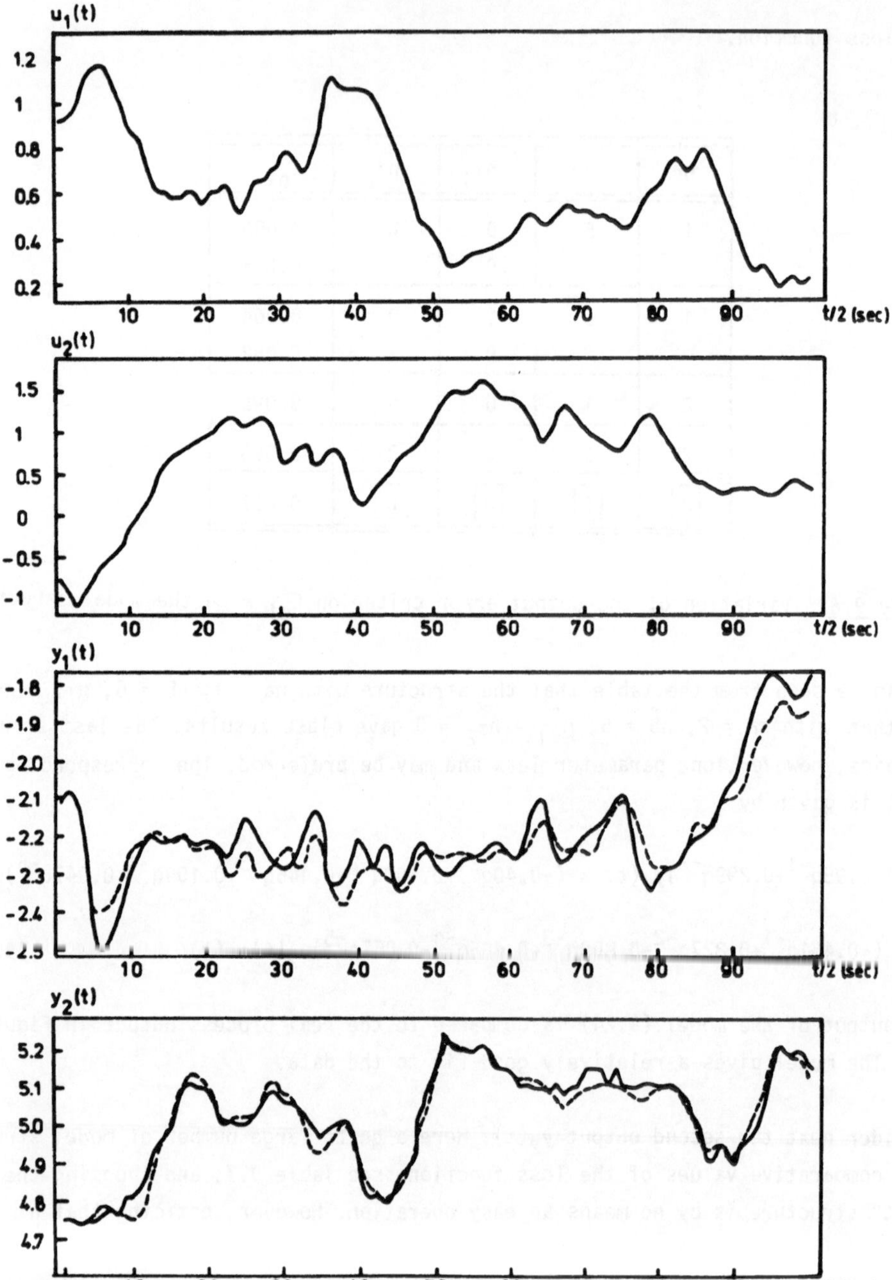

Figure 9.7. Measured inputs and outputs and computed model outputs for the turbo-alternator. The model (9.24), (9.26) was used. The mean values are $\bar{y}_1 = -2.19$, $\bar{y}_2 = 5.02$, $\bar{u}_1 = 0.61$, $\bar{u}_2 = 0.68$.

na	nb	n_{τ_1}	n_{τ_2}	C_{o2}
1	1	0	0	0.050
		1	1	0.048
		2	0	0.048
		2	1	0.046
1	2	1	1	0.047
1	3	0	0	0.046
1	4	0	0	0.042
		1	1	0.038
1	5	0	0	0.038
2	1	1	1	0.039
		2	0	0.045
		2	1	0.052
2	2	2	0	0.037
		2	3	0.036
2	5	1	1	0.036
2	6	0	0	0.033
3	1	0	0	0.040
		2	1	0.043
3	2	2	2	0.044
		2	3	0.032
3	3	0	0	0.052
		1	1	0.030
3	4	0	0	0.030
3	6	0	0	0.037
4	1	0	0	0.040
		1	1	0.034
5	1	1	1	0.034
6	1	1	1	0.033
6	2	0	0	0.035

Table 9.7. **Variation of the output error criterion C_{o2} with the model structure.**

gives quite a good fit to the data, cf Figure 9.7.

We may remark that the simpler structure

$$na = 2 \qquad nb = 1 \qquad n\tau_1 = n\tau_2 = 1 \tag{9.27}$$

might seem a good alternative to (9.25). However, if the "true" structure would be (9.27), then the model structure (9.25) should not be identifiable, see the discussion in Section 9.4, and as a consequence the corresponding loss function C_{02} should have a large value. Since (9.25) gave the smallest value of C_{02} among the structures tested, (9.27) should not be preferred to (9.25). Note, however, that the model structure na = 3, nb = 2, $n\tau_1 = 2$, $n\tau_2 = 3$ might be a good alternative to (9.25).

Finally, the estimated model (9.24), (9.26) will be compared to some extent to the model of the turbo-alternator obtained by Jenkins and Watts (1969). A direct and complete comparison would be somewhat difficult since the models are of different types. However, the gains of the various elements of the transfer function matrices corresponding to the models in question can easily be compared.

Jenkins and Watts's model has the following static gain matrix

$$\hat{G}(1) = \begin{bmatrix} -0.55 & -0.22 \\ -0.26 & 0.04 \end{bmatrix}$$ [1)]

while for our model

$$\hat{G}(1) = \begin{bmatrix} -0.54 & -0.24 \\ -0.27 & 0.07 \end{bmatrix}$$

The static gains of the two models are clearly quite similar.

---

1) The minus signs do not appear in Jenkins and Watts (1969). However, it is known that an increase of the in-phase current leads to decreases in the amplitude and frequency of the voltage and also that the voltage amplitude decreases when the out-of-phase current increases.

APPENDICES

Appendix 1

PERSISTENTLY EXCITING SIGNALS

In this appendix we will review some results on persistent excitation.

Definition A1.1. A signal u(t) is said to be persistently exciting (p.e.) of order n if the limits

$$m = \lim_{N\to\infty} \frac{1}{N}\sum_{t=1}^{N} u(t) \tag{A1.1a}$$

$$r_u(\tau) = \lim_{N\to\infty} \frac{1}{N}\sum_{t=1}^{N} [u(t)-m][u(t+\tau)-m]^T \qquad (\text{all } \tau) \tag{A1.1b}$$

exist with probability one, and the matrix

$$R_n = \begin{bmatrix} r_u(0) & \dots & r_u(1-n) \\ \vdots & \ddots & \vdots \\ r_u(n-1) & \dots & r_u(0) \end{bmatrix} \tag{A1.2}$$

is positive definite. ∎

A stationary ergodic process u(t) is thus p.e. of order n if and only if the associated covariance matrix

$$R_n = E\begin{bmatrix} u(t-1)-m \\ \vdots \\ u(t-n)-m \end{bmatrix}[u^T(t-1)-m^T \dots u^T(t-n)-m^T] \tag{A1.3}$$

is positive definite. Here $m = Eu(t)$.

If u(t) is p.e. of order n it follows in particular that the second order moment matrix

$$E\begin{bmatrix} u(t-1) \\ \vdots \\ u(t-n) \end{bmatrix}[u^T(t-1)\dots u^T(t-n)] = R_n + \begin{bmatrix} m \\ \vdots \\ m \end{bmatrix}[m^T \dots m^T]$$

is positive definite.

In what follows we will review some properties of persistently exciting signals. Such an analysis was originally undertaken by Ljung (1971). Extension of some results to multivariable signals was later considered by Stoica (1981 b). To simplify the proofs we restrict them to stationary ergodic, stochastic processes, but similar results hold for deterministic signals too, see Ljung (1971) for more details.

Result A1.1. Let u(t) be a multivariable ergodic process. Assume that its spectral density matrix is positive definite in at least n distinct frequencies (within the interval $(-\pi, \pi)$). Then u(t) is persistently exciting of order n.

Proof. Let $g = [g_1^T \dots g_n^T]^T$ be an arbitrary n·nu-vector and set $G(q^{-1}) = \sum_1^n g_i q^{-i}$. Consider then the equation

$$0 = g^T R_n g = \frac{1}{2\pi} \int_{-\pi}^{\pi} G^T(e^{i\omega}) \phi_u(\omega) G(e^{-i\omega}) d\omega$$

where $\phi_u(\omega)$ is the spectral density matrix of u(t). Since it is nonnegative definite we get

$$G^T(e^{i\omega}) \phi_u(\omega) G(e^{-i\omega}) \equiv 0$$

We can then conclude that $G(e^{-i\omega})$ is zero in n distinct frequencies. However, as G(z) is a (vector) polynomial of degree n-1 only, this implies g = 0. Thus the matrix R_n is positive definite and u(t) is persistently exciting of order n.

■

Result A1.2. An ARMA process is persistently exciting of any order.

Proof. The assertion follows immediately from Result A1.1 since the spectral density matrix of an ARMA process is positive definite for almost all frequencies in $(-\pi, \pi)$. ■

For *scalar* processes the condition of Result A1.1 is also necessary for u(t) to be p.e. of order n, see Result A1.3 below. This is not true in the multivariable case, as shown by Stoica (1981 b).

Result A1.3. Let u(t) be a scalar signal that is persistently exciting of order n. Then its spectral density is nonzero in at least n frequencies.

Proof. The proof is by contradiction. Using the calculations of Result A1.1 we have

$$0 = g^T R_n g \Leftrightarrow \phi_u(\omega)|G(e^{i\omega})|^2 \equiv 0$$

Assume that the spectral density is nonzero in at most n-1 frequencies. Then we can choose the polynomial G(z) (of degree n-1) to vanish where $\phi_u(\omega)$ is nonzero. This means that there is a nonzero vector g such that $g^T R_n g = 0$. Hence u is not p.e. of order n. This is a contradiction and the result is proved. ∎

Result A1.4. Let u(t) be a multivariable ergodic signal with spectral density matrix $\phi_u(\omega)$. Assume that $\phi_u(\omega)$ is positive definite for at least n distinct frequencies. Let $H(q^{-1})$ be a (multivariable) asymptotically stable rational filter and assume that det[H(z)] has no zero on the unit circle. Then the filtered signal $y(t) = H(q^{-1})u(t)$ is persistently exciting of order n.

Proof. Since

$$\phi_y(\omega) = H(e^{i\omega})\phi_u(\omega)H^T(e^{-i\omega})$$

the result is immediate from Result A1.1. ∎

The above result can be somewhat strengthened for scalar signals.

Result A1.5. Let u(t) be a scalar signal that is persistently exciting of order n. Assume that $H(q^{-1})$ is an asymptotically stable filter with k zeros on the unit circle. Then the filtered signal $y(t) = H(q^{-1})u(t)$ is persistently exciting of order m with $n-k \le m \le n$.

Proof. Since we have

$$\phi_y(\omega) = \phi_u(\omega)|H(e^{i\omega})|^2$$

the result follows from Results A1.1, A1.3. ∎

Note in particular that if $H(q^{-1})$ has no zeros on the unit circle then u(t) and $H(q^{-1})u(t)$ are p.e. of the same order. This can also be seen as a consequence of Lemma A3.7.

We turn now to a special class of vector-valued signals whose components are different powers of one same scalar signal. Such vector signals appear in the study of Hammerstein systems. We will say that the scalar signal is *strongly persistently exciting* if and only if the corresponding vector signal is p.e. For a formal definition see below.

Definition A1.2. The stationary signal $\bar{u}(t)$ is said to be strongly persistently exciting or orders n, m [denoted spe(n, m)] if and only if the matrix

$$R_u(n,m) = E\begin{bmatrix} u(t-1)-m_u \\ \vdots \\ u(t-n)-m_u \end{bmatrix}[u^T(t-1)-m_u^T \dots\ u^T(t-n)-m_u^T] \tag{A1.4}$$

is positive definite. The vector u(t) of dimension m is given by

$$u(t) = [\bar{u}(t)\ \bar{u}^2(t) \dots\ \bar{u}^m(t)]^T$$

and $m_u = Eu(t)$. ∎

We now proceed to show that ARMA processes are strongly persistently exciting signals of any finite orders.

Lemma A1.1. Assume that

$$\bar{u}(t) = H(q^{-1})e(t) \tag{A1.5}$$

where $H(q^{-1})$ is an asymptotically stable filter with H(0) = 1, and e(t), t = 0, ±1,... is a sequence of independent and identically distributed random variables with zero mean and finite moments. Assume that the distribution of e(t) is nonzero in at least m+1 distinct points. Then $\bar{u}(t)$ is spe(n,m) for any finite n and m.

Proof. Introduce the notations

$$\bar{v}(t) = \begin{bmatrix} u(t-1) \\ \vdots \\ u(t-n) \end{bmatrix} \qquad m_{\bar{v}} = E\bar{v}(t) = \begin{bmatrix} m_u \\ \vdots \\ m_u \end{bmatrix}$$

Consider now the equation

$$\bar{a}^T R_u(n,m)\bar{a} = \bar{a}^T E[\bar{v}(t)-m_{\bar{v}}][\bar{v}(t)-m_{\bar{v}}]^T\bar{a} = 0 \tag{A1.6}$$

where the unknown vector is $\bar{a} = [\bar{a}_1^T\ \bar{a}_2^T \ldots \bar{a}_n^T]^T$. (A1.6) can be written as

$$\sum_{i=1}^{n} \bar{a}_i^T[u(t-i)-m_u] = 0 \quad \text{w.p.1} \tag{A1.7}$$

Define

$$w(t) = \begin{bmatrix} e(t) \\ \vdots \\ e^m(t) \end{bmatrix} \qquad m_w = Ew(t)$$

Since $e(t)$ and $e(s)$ are independent for $t \neq s$ we also have $e(t)$, $u(s)$ independent for $t > s$. By post-multiplying (A1.7) with $[w(t-1)-m_w]^T$ and taking expectation we then get

$$\bar{a}_1^T E[u(t-1)-m_u][w(t-1)-m_w]^T = 0 \tag{A1.8}$$

Now decompose $\bar{u}(t)$ as $\bar{u}(t) = e(t)+\varepsilon(t)$, where $e(t)$ and $\varepsilon(t)$ clearly are independent. We then get

$$u(t) = \begin{bmatrix} \varepsilon(t) \\ \vdots \\ \varepsilon^m(t) \end{bmatrix} + \begin{bmatrix} 1 & & \bigcirc \\ s_{21} & 1 & \\ \vdots & & \ddots \\ s_{m1} & \cdots & & 1 \end{bmatrix} \begin{bmatrix} e(t) \\ \vdots \\ e^m(t) \end{bmatrix} \triangleq \bar{\varepsilon}(t)+S(t)w(t) \tag{A1.9}$$

The elements of $S(t)$ are functions of $\varepsilon(t)$ but these are independent of $w(t)$. To be specific we have

$$s_{i+k,i} = \frac{(i+k)!}{k!\,i!}\varepsilon^k(t) \qquad i = 1,\ldots, m \qquad k = 0,\ldots, m-i$$

Using (A1.9) in (A1.8) one obtains

$$0 = a_1^T E[\bar{\varepsilon}(t-1)+S(t-1)w(t-1)-m_u][w(t-1)-m_w]^T$$
$$= a_1^T ES(t-1)\cdot E[w(t-1)-m_w][w(t-1)-m_w]^T \tag{A1.10}$$

Note that by construction ES(t) is nonsingular. Our present aim is to show that $a_1 = 0$. It is then sufficient to consider the equation

$$0 = \alpha^T E[w(t)-m_w][w(t)-m_w]^T\alpha = E[\alpha^T\{w(t)-m_w\}]^2 \tag{A1.11}$$

where $\alpha = [\alpha_1 \dots \alpha_m]^T$. Let $f_e(\cdot)$ be the probability density function of e(t). Then (A1.11) gives

$$0 = E[\sum_{i=1}^{m} \alpha_i\{e^i(t)-m_{wi}\}]^2 = \int_{-\infty}^{\infty}[\sum_{i=1}^{m} \alpha_i(z^i-m_{wi})]^2 f_e(z)dz \tag{A1.12}$$

Let $\{z_j\}_{j=1}^{m+1}$ be distinct points where $f_e(z)$ is strictly positive. Then according to (A1.12) we must have

$$\sum_{i=1}^{m} \alpha_i(z_j^i-m_{wi}) = 0 \qquad j = 1,\dots, m+1$$

However, since the left hand side is a polynomial in z of degree m it follows that $\alpha = 0$. This means that the matrix $E[w(t)-m_w][w(t)-m_w]^T$ is positive definite and hence (A1.10) implies $a_1 = 0$. By repeating the reasoning, it follows from (A1.7) that $\bar{a} = 0$. Thus the large matrix $E[\bar{v}(t)-m_{\bar{v}}][\bar{v}(t)-m_{\bar{v}}]^T$ is positive definite, which proves that u(t) is spe(n,m). ■

Appendix 2

SOME RESULTS ON POLYNOMIALS AND ANALYTIC FUNCTIONS

We first give a result on reciprocal polynomials.

Consider the following two polynomials

$$A(z) = A_0+A_1z+\ldots+A_{na}z^{na}$$

$$B(z) = B_0+B_1z+\ldots+B_{nb}z^{nb}$$

of dimensions ny|nw and ny|nu, respectively. Introduce the corresponding reciprocal polynomials

$$A^*(z) = z^{na}A(z^{-1})$$

$$B^*(z) = z^{nb}B(z^{-1})$$

As is known, A(z), B(z) left coprime does not necessarily imply that A*(z), B*(z) are also left coprime. We have, however, the following simple result.

Lemma A2.1. The following statements are equivalent.

i) A(z), B(z) are left coprime and rank $[A_{na}\ B_{nb}]$ = ny.

ii) A*(z), B*(z) are left coprime and rank $[A_0\ B_0]$ = ny.

Proof. Assume that i) holds. Then we have, see Kailath (1980),

$$\text{rank}[A(z)\ B(z)] = ny \qquad \forall z \in C \tag{A2.1}$$

It follows from (A2.1) that

$$\text{rank}[A_0\ B_0] = ny.$$

Furthermore, for $z \neq 0$ we can write

$$\text{rank } [A^*(z)\ B^*(z)] = \text{rank } [z^{na}A(z^{-1})\ z^{nb}B(z^{-1})] = \text{rank } [A(z)\ B(z)] = ny$$

which together with rank $[A_{na}\ B_{nb}] = ny$ shows that $A^*(z)$ and $B^*(z)$ are left coprime. The implication i) ⇒ ii) is thus proved. The implication ii) ⇒ i) follows similarly since $A(z) = [A^*(z)]^*$. ■

We next proceed to give some results on analytic functions. The first is taken from Åström and Söderström (1974).

Lemma A2.2. Consider the function

$$f(z) = \frac{g(z)}{\prod_{i=1}^{k} (z-u_i)^{t_i}} \tag{A2.2}$$

where $g(z)$ is analytic inside and on the unit circle, the complex numbers $\{u_i\}$ fulfil $|u_i| < 1$ and are distinct, and $t_i \geq 1$ are some integers. Assume that

$$\frac{1}{2\pi i} \oint f(z) z^j \frac{dz}{z} = 0 \qquad j = 1, \ldots, m \tag{A2.3}$$

where the integration path is the unit circle and

$$\sum_{i=1}^{k} t_i \leq m \tag{A2.4}$$

Then $f(z)$ is analytic inside the unit circle, i.e. the polynomial $\prod_{i=1}^{k} (z-u_i)^{t_i}$ divides the function $g(z)$.

Proof. Equation (A2.3) can with use of (A2.2) be rewritten as ($D^{(\nu)}$ denoting the ν:th derivative)

$$0 = \frac{1}{2\pi i} \oint \frac{g(z)}{\prod_{i=1}^{k} (z-u_i)^{t_i}} z^{j-1} dz = \sum_{\ell=1}^{k} \operatorname*{res}_{z=u_\ell} \frac{g(z)}{\prod_{i=1}^{k} (z-u_i)^{t_i}} z^{j-1}$$

$$= \sum_{\ell=1}^{k} \frac{1}{(t_\ell - 1)!} D^{(t_\ell - 1)} \left[\frac{g(z)}{\prod_{\substack{i=1 \\ i \neq \ell}}^{k} (z-u_i)^{t_i}} \cdot z^{j-1} \right]_{z=u_\ell}$$

$$= \sum_{\ell=1}^{k} \frac{1}{(t_\ell-1)!} \sum_{\nu=0}^{t_\ell-1} \binom{t_\ell-1}{\nu} D^{(\nu)}[z^{j-1}]_{|z=u_\ell}$$

$$\cdot\ D^{(t_\ell-1-\nu)}\left[\frac{g(z)}{\prod_{\substack{i=1\\ i\neq\ell}}^{k} (z-u_i)^{t_i}}\right]_{|z=u_\ell} \qquad j = 1,\ldots, m$$

In matrix formulation these equations become

$$Ud = [U_1 \ldots U_k] \begin{bmatrix} d_1 \\ \vdots \\ d_k \end{bmatrix} = 0 \tag{A2.5}$$

where the block $U_\ell (\ell = 1,\ldots, k)$ is a $m|t_\ell$ matrix given by

$$U_\ell = \begin{bmatrix} 1 & 0 & \ldots & 0 \\ z & 1 & & \\ \vdots & & & \\ z^{m-1} & D[z^{m-1}] & \ldots & D^{(t_\ell-1)}[z^{m-1}] \end{bmatrix}_{|z=u_\ell}$$

The vector $d_\ell (\ell = 1,\ldots, k)$ has t_ℓ components and is given by

$$d_{\ell,j} = \frac{1}{(j-1)!(t_\ell-j)!} D^{(t_\ell-j)}\left[\frac{g(z)}{\prod_{\substack{i=1\\ i\neq\ell}}^{k} (z-u_i)^{t_i}}\right]_{|z=u_\ell} \qquad j = 1,\ldots, t_\ell$$

The matrix U is a generalized Vandermonde matrix. It follows e.g. from Kaufman (1969), that if (A2.4) holds then rank $U = \sum_{i=1}^{k} t_i$. Thus, it is concluded from (A2.5) that $d = 0$, or

$$D^{(t_\ell-j)}\left[\frac{g(z)}{\prod_{\substack{i=1\\ i\neq\ell}}^{k} (z-u_i)^{t_i}}\right]_{|z=u_\ell} = 0 \qquad j = 1,\ldots, t_\ell, \quad \ell = 1,\ldots, k$$

This relation can be written as

$$\sum_{\nu=0}^{t_\ell-j} \binom{t_\ell-j}{\nu} D^{(\nu)}[g(z)]_{|z=u_\ell}$$

$$\cdot\, D^{(t_\ell - j - \nu)}\left[\frac{1}{\prod_{\substack{i=1\\ i\neq \ell}}^{k}(z-u_i)^{t_i}}\right]_{|z=u_\ell} = 0 \qquad j = 1,\ldots, t_\ell, \quad \ell = 1,\ldots, k \tag{A2.6}$$

Consider first (A2.6) for $j = t_\ell$. Then $D^{(0)}[g(z)]_{|z=u_\ell} = 0$ is easily derived. Consider next (A2.6) for $j = t_\ell - 1$, which in a similar way implies $D^{(1)}[g(z)]_{|z=u_\ell} = 0$. By repeated use of this argument it can be concluded that

$$D^{(\nu)}[g(z)]_{|z=u_\ell} = 0, \qquad \nu = 0,\ldots, t_\ell - 1, \qquad \ell = 1,\ldots, k \tag{A2.7}$$

Equation (A2.7) implies that the polynomial $\prod_{i=1}^{k}(z-u_i)^{t_i}$ divides the function g(z). This means that all poles of the function f inside the unit circle are cancelled. With this observation the proof of the lemma is finished. ∎

In the next lemma we need the concept of Lebesgue measure. For our purposes it is sufficient to define it in the following way. Let Ω be a subset of R^n. Then its Lebesgue measure is

$$\mu(\Omega) = \int_\Omega 1\, dx_1 \ldots dx_n$$

For $n = 2$ $\mu(\Omega)$ can be interpreted as the area and for $n = 3$ as the volume of Ω. Note in particular that if Ω is a subset of lower dimension than n (such as a line for $n = 2$ or a plane for $n = 3$) then it will have zero Lebesgue measure. For a more stringent discussion on the Lebesgue measure, see for example Cramér (1946) or Pearson (1974).

Lemma A2.3. Let $\Omega \subset R^n$ be an open connected set. Consider a function $f: \Omega \to R$ such that

i) $f(x)$ is analytic in x_i $(i = 1,\ldots, n)$ for $x \in \Omega$

ii) there is a point $x^* \in \Omega$ such that $f(x^*) \neq 0$

Then the set

$$M = \{x \mid x \in \Omega,\ f(x) = 0\}$$

has zero Lebesgue measure.

Proof. The proof is by contradiction. Assume that M has a strictly positive measure. Then there exists a point $\bar{x}$ and an open neighbourhood $B \subset \Omega$ of $\bar{x}$ such that $f(x) = 0$ for all $x \in B$. It then follows from the uniqueness theorem for analytic functions (which is the basis for analytic continuation) that $f(x) \equiv 0$ in Ω. However, this is a contradiction to $f(x^*) \neq 0$. ■

Corollary. Let $\Omega \subset R^n$ be an open connected set. Consider a $m|p$ dimensional matrix $R(\rho)$, $m \geq p$ where the vector $\rho \in \Omega$. Assume that

i) each element of $R(\rho)$ is an analytic function in ρ_i, $i = 1,\dots, n$. (for $\rho \in \Omega$)

ii) there exists a $\rho^* \in \Omega$ such that rank $R(\rho^*) = p$.

Then the set

$$M = \{\rho | \rho \in \Omega, \text{ rank } R(\rho) < p\}$$

has Lebesgue measure zero.

Proof. Take $f(\rho) = \det [R^T(\rho)R(\rho)]$. Since $f(\rho) \neq 0$ is equivalent to rank $R(\rho) = p$ the corollary follows from Lemma A2.3. ■

Appendix 3

SOME MATRIX RESULTS

We first consider Sylvester matrices. This is a topic dealt with by Jury (1974), Anderson and Jury (1976), Kailath (1980) etc. Such matrices can be defined in some different ways. The way chosen here is convenient for the analysis. The determinant of a square Sylvester matrix is called "the resultant". Sylvester matrices are always associated with two polynomials. We start with the case when both the polynomials are scalar.

<u>*Definition A3.1*</u>. Consider two polynomials

$$\begin{aligned} A(z) &= a_0 z^{na} + a_1 z^{na-1} + \dots + a_{na} \\ B(z) &= b_0 z^{nb} + b_1 z^{nb-1} + \dots + b_{nb} \end{aligned} \tag{A3.1}$$

Then the Sylvester matrix of dimension $(\overline{na}+\overline{nb})|(\overline{na}+\overline{nb})$ with

$$\min(\overline{na}-na,\ \overline{nb}-nb) = 0 \tag{A3.2}$$

is defined as

$$S(A,B) = \left[\begin{array}{ccccccc} a_0 & a_1 & \dots & a_{na} & & & \bigcirc \\ & \ddots & & & \ddots & & \\ \bigcirc & & a_0 & a_1 & \dots & & a_{na} \\ \hline b_0 & b_1 & \dots & b_{nb} & & & \bigcirc \\ & \ddots & & & \ddots & & \\ \bigcirc & & b_0 & b_1 & \dots & & b_{nb} \end{array}\right] \begin{array}{l} \\ \overline{nb} \text{ rows} \\ \\ \\ \overline{na} \text{ rows} \\ \\ \end{array} \tag{A3.3}$$

■

<u>*Lemma A3.1*</u>. Consider the Sylvester matrix $S(A,B)$, (A3.3). Assume that $A(z)$ and $B(z)$ have exactly k common zeros. Then

$$\operatorname{rank} S(A,B) = \overline{na}+\overline{nb}-k \tag{A3.4}$$

Proof. Consider the equation

$$x^T S(A,B) = 0 \tag{A3.5}$$

where

$$x^T = [\tilde{b}_1 \dots \tilde{b}_{\overline{nb}}\ \tilde{a}_1 \dots \tilde{a}_{\overline{na}}]$$

and let

$$\tilde{A}(z) = \tilde{a}_1 z^{\overline{na}-1} + \dots + \tilde{a}_{\overline{na}} \qquad \tilde{B}(z) = \tilde{b}_1 z^{\overline{nb}-1} + \dots + \tilde{b}_{\overline{nb}}$$

Then (A3.5) can be written as

$$\tilde{B}(z)A(z) + \tilde{A}(z)B(z) = 0 \tag{A3.6}$$

Due to the assumption we can write

$$A(z) = A_0(z)L(z) \qquad B(z) = B_0(z)L(z)$$

$$L(z) = z^k + \ell_1 z^{k-1} + \dots + \ell_k$$

$A_0(z)$, $B_0(z)$ coprime, $\deg A_0 = na-k$, $\deg B_0 = nb-k$

Thus (A3.6) is equivalent to

$$\tilde{B}(z)A_0(z) = -\tilde{A}(z)B_0(z)$$

Since both sides must have the same zeros it follows that the general solution can be written as

$$\tilde{A}(z) = A_0(z)M(z)$$

$$\tilde{B}(z) = -B_0(z)M(z)$$

where

$$M(z) = m_1 z^{k-1} + \dots + m_k$$

has arbitrary coefficients. This means that x lies in a k dimensional subspace. However, this subspace is $N(S^T(A,B))$, cf (A3.5), and thus its dimension must be equal to $\overline{na}+\overline{nb}$-rank $S(A,B)$ which proves the lemma. If $k = 0$ we have $M(z) = 0$ and only the trivial solution $x = 0$ to (A3.5) exists. ■

Remark. Note that the proof has also given a characterization of the nullspace of $S^T(A,B)$. ■

Corollary A3.1. $A(z)$ and $B(z)$ are coprime if and only if $S(A,B)$ is nonsingular. ■

We now extend the result to the case when the polynomial $A(z)$ is a vector polynomial.

Definition A3.2. Consider the polynomials

$$A(z) = A_0 z^{na}+\ldots+A_{na}$$
$$B(z) = b_0 z^{nb}+\ldots+b_{nb} \qquad \text{(A3.7)}$$

the coefficients A_i being $1|m$ vectors. The generalized Sylvester matrix of dimension $(m\cdot\overline{na}+\overline{nb})|(m(\overline{na}+\overline{nb}))$ with

$$\min(\overline{na}-na,\ \overline{nb}-nb) = 0 \qquad \text{(A3.8)}$$

is defined as

$$\bar{S}(A,B) = \left[\begin{array}{ccccccc} A_0 & A_1 & \ldots & A_{na} & & & 0 \\ & \ddots & & & \ddots & & \\ 0 & & A_0 & A_1 & \ldots & A_{na} & \\ \hline b_0I & b_1I & \ldots & b_{nb}I & & & 0 \\ & \ddots & & & \ddots & & \\ 0 & & b_0I & b_1I & \ldots & & b_{nb}I \end{array}\right] \begin{array}{l} \overline{nb} \text{ rows} \\ \\ m\cdot\overline{na} \text{ rows} \end{array} \qquad \text{(A3.9)}$$

The identity matrices in (A3.9) have all dimension $m|m$. ■

Remark. Note that for $m = 1$ the former Sylvester matrix (A3.3) is obtained. Also note that for $m > 1$ the generalized Sylvester matrix will be rectangular, with more columns than rows. ■

Lemma A3.2. Consider the generalized Sylvester matrix $\bar{S}(A,B)$, (A3.9). Assume that A(z) and B(z) have exactly k common zeros [1]. Then

$$\text{rank } \bar{S}(A,B) = m\cdot\overline{na}+\overline{nb}-k \tag{A3.10}$$

Proof. The proof parallels that of Lemma A3.1. One has just to remember that A(z) is (and let $\tilde{A}(z)$ be) a vector polynomial. ∎

Corollary A3.2. A(z) and B(z) are coprime if and only if $\bar{S}(A,B)$ has full rank equal to $m\cdot\overline{na}+\overline{nb}$. ∎

We next give some lemmas that concern certain covariance matrices. We are interested in finding conditions for nonsingularity and, in some cases, also the eigenvalue location.

Lemma A3.3. Let $\varphi(t)$ be a stochastic vector such that $E\varphi(t)\varphi^T(t)$ is positive definite. Assume that the scalar filter $H(q^{-1})$ is strictly positive real [i.e. $\text{Re } H(e^{i\omega}) > 0 \quad -\pi \le \omega \le \pi$]. Then the matrix

$$P = E\varphi(t)\cdot H(q^{-1})\varphi^T(t) \tag{A3.11}$$

is positive definite.[2]

Proof. Let h be an arbitrary vector of the same dimension as $\varphi(t)$ and put $v(t) = \varphi^T(t)h$. Consider the following calculation

$$h^TPh = E\{h^T\varphi(t)\}\cdot H(q^{-1})\{\varphi^T(t)h\} = Ev(t)\cdot H(q^{-1})v(t)$$

$$= \frac{1}{2\pi}\int_{-\pi}^{\pi} H(e^{i\omega})\phi_v(\omega)d\omega = \frac{1}{2\pi}\int_{-\pi}^{\pi} [\text{Re } H(e^{i\omega})]\phi_v(\omega)d\omega \ge 0$$

Here, $\phi_v(\omega)$ denotes the spectral density of v(t).

---

1) A complex number $\tilde{z}$ is said to be a zero of the vector polynomial A(z) if $A(\tilde{z}) = 0$.

2) We use the common convention that a (possibly nonsymmetric) matrix P is positive definite if $x \neq 0$ implies $x^TPx > 0$.

Clearly, the calculation shows that P is at least nonnegative definite. Moreover $h^TPh = 0$ implies $\phi_v(\omega) \equiv 0$ or $v(t) \equiv 0$ w.p.1. Due to the assumptions on $\varphi(t)$ this gives $h = 0$, which proves that P is positive definite. ∎

Lemma A3.4. Let $\varphi(t)$ be a stochastic vector such that

$$Q = E\varphi(t)\varphi^T(t) \tag{A3.12}$$

is positive definite. Let $H(q^{-1})$ be a scalar strictly positive real filter and define

$$S = E\varphi(t)\cdot H(q^{-1})\varphi^T(t) \tag{A3.13}$$

Then all eigenvalues of $Q^{-1}S$ will have strictly positive real parts.

Proof. Since the inverse Q^{-1} exists there is a nonsingular matrix L such that $Q^{-1} = LL^T$. The matrix $Q^{-1}S$ has the same eigenvalues as

$$L^{-1}Q^{-1}SL = L^TSL$$

Let $\lambda = \lambda_R + i\lambda_I$ be an arbitrary eigenvalue of L^TSL with the associated eigenvector $\mu = \mu_R + i\mu_I$ ($\lambda_R, \lambda_I, \mu_R, \mu_I$ being real-valued quantities). Since L^TSL is a real-valued matrix we have

$$L^TSL\mu_R = \lambda_R\mu_R - \lambda_I\mu_I, \qquad L^TSL\mu_I = \lambda_R\mu_I + \lambda_I\mu_R$$

The real part λ_R can then be found to satisfy

$$\lambda_R = \frac{\mu_R^TL^TSL\mu_R + \mu_I^TL^TSL\mu_I}{\mu_R^T\mu_R + \mu_I^T\mu_I} > 0$$

where the inequality follows from Lemma A3.3. ∎

Remark. An alternative proof based on properties of Lyapunov equations has been given by Ljung (1977 b). ∎

For certain $\varphi(t)$ vectors it is even possible with modest efforts to determine the eigenvalues of $Q^{-1}S$ explicitly, see Lemma A3.5 below.

Lemma A3.5. Let

$$\varphi(t) = \frac{1}{F(q^{-1})} [e(t-1) \dots e(t-m)]^T \tag{A3.14}$$

with e(t) being white noise of zero mean and variance σ, and $F(q^{-1})$ a polynomial in the unit delay operator q^{-1}, of degree less than or equal to m.

$$F(q^{-1}) = 1+f_1q^{-1}+\dots+f_mq^{-m} \qquad \text{(possibly, } f_m = 0 \text{ etc.)}$$

Assume that F(z), with z a complex variable replacing q^{-1}, has all zeros outside the unit circle. Define

$$M = [E\varphi(t)\varphi^T(t)]^{-1}[E\varphi(t)\cdot H(q^{-1})\varphi^T(t)] \tag{A3.15}$$

where $H(q^{-1})$ is a rational, asymptotically stable, scalar-valued filter.

Then the eigenvalues of the matrix M are given by

$$H(\alpha_i) \qquad i = 1,\dots, m$$

where $\{\alpha_i\}_{i=1}^m$ are the zeros of the reciprocal polynomial

$$F^*(z) \triangleq z^mF(z^{-1})$$

Proof. We shall first assume that F*(z) has m distinct zeros. Then we shall treat the general case of multiple zeros using a perturbation argument.

Therefore, assume first that F*(z) has distinct zeros $\{\alpha_i\}_{i=1}^m$. The eigenvalues of M are given by

$$\det[\lambda I-M] = 0$$

or equivalently

$$\det(\Lambda) = 0 \quad \text{with} \quad \Lambda = E[\lambda-H(q^{-1})]\varphi(t)\cdot\varphi^T(t) \tag{A3.16}$$

Introduce the residuals of $\frac{1}{F(z)F^*(z)}$ in the zeros of F*(z)

$$\beta_k \triangleq \lim_{z\to\alpha_k} \frac{z-\alpha_k}{F(z)F^*(z)} \qquad k = 1,\dots, m$$

Then we can write

$$\Lambda = \frac{\sigma}{2\pi i}\oint[\lambda - H(z)]\,\frac{1}{F(z)F^*(z)}\begin{bmatrix}1\\ \vdots\\ z^{m-1}\end{bmatrix}[z^{m-1}\ldots\ 1]dz$$

$$= \sigma\sum_{k=1}^{m}\beta_k[\lambda - H(\alpha_k)]\begin{bmatrix}1\\ \alpha_k\\ \vdots\\ \alpha_k^{m-1}\end{bmatrix}[\alpha_k^{m-1}\ldots\ \alpha_k\ 1]$$

$$= \sigma\begin{bmatrix}1 & \ldots & 1\\ \alpha_1 & & \alpha_m\\ \vdots & & \vdots\\ \alpha_1^{m-1} & \ldots & \alpha_m^{m-1}\end{bmatrix}\begin{bmatrix}\beta_1\{\lambda - H(\alpha_1)\} & & \bigcirc\\ & \ddots & \\ \bigcirc & & \beta_m\{\lambda - H(\alpha_m)\}\end{bmatrix}\begin{bmatrix}\alpha_1^{m-1} & \ldots & \alpha_1 & 1\\ \vdots & & & \vdots\\ \alpha_m^{m-1} & \ldots & \alpha_m & 1\end{bmatrix} \qquad (A3.17)$$

Now the first matrix in (A3.17) as well as the third are Vandermonde matrices associated with $F^*(z)$. Since $\{\alpha_i\}$ were assumed to be distinct, these matrices are nonsingular. Furthermore, $\beta_k \neq 0$, $k = 1,\ldots, m$ due to the assumption of distinct zeros. Then it follows easily from (A3.16) and (A3.17) that the eigenvalues of M are given by $\{H(\alpha_k)\}_{k=1}^{m}$.

Assume now that the polynomial $F^*(z)$ has multiple zeros. Let $F_\varepsilon^*(z)$ be a polynomial with distinct zeros located inside the unit circle, and such that the coefficients of $F^*(z)-F_\varepsilon^*(z)$ have magnitudes less than a positive number ε. Such a polynomial can always be found no matter how small ε is taken. Denote the matrix (A3.15) corresponding to $F_\varepsilon^*(z)$ by M_ε. According to the above analysis, the eigenvalues of M_ε are $\{H(\alpha_k^\varepsilon)\}$, where $\{\alpha_k^\varepsilon\}$ denote the zeros of $F_\varepsilon^*(z)$.

However, the eigenvalues are continuous functions of the matrix elements, which in turn vary continuously with the coefficients of $F^*(z)$. It thus follows that the eigenvalues of M are given by

$$\{\lambda_k(M)\}_{k=1}^{m} = \{\lim_{\varepsilon\to 0}\lambda_k(M_\varepsilon)\}_{k=1}^{m} = \{\lim_{\varepsilon\to 0} H(\alpha_k^\varepsilon)\}_{k=1}^{m} = \{H(\alpha_k)\}_{k=1}^{m}$$

The proof is thus complete. ∎

Remark. In Stoica et al (1981) we have extended the above lemma to the case of *matrix*-valued H-filters. The result becomes in such a case quite involved and we have chosen not to include it here. It is, however, worth noting that for matrix-valued H the perturbational analysis used in the above proof to handle the case of multiple zeros does not seem to be readily applicable. In Stoica et al (1981) we have therefore approached the case of multiple zeros directly. ■

We continue with giving the matrix inversion lemma.

Lemma A3.6. Let A, B, C, D be matrices of compatible dimensions. Assuming the inverses appearing below exist, then

$$[A+BCD]^{-1} = A^{-1}-A^{-1}B[C^{-1}+DA^{-1}B]^{-1}DA^{-1} \tag{A3.18}$$

Proof. By direct multiplication of the right hand side with (A+BCD). For some remarks on the origin of this result, see Kailath (1980). ■

The following three lemmas all deal with covariance matrices.

Lemma A3.7. Let $F(q^{-1})$ be an ny|ny-dimensional asymptotically stable filter. Assume that $\det\{F(q^{-1})\}$ has no zero on the unit circle. Let $\tilde{\phi}^T(t)$ be a stationary stochastic matrix of dimension ny|nθ. Then the matrices

$$R_F = E[F(q^{-1})\tilde{\phi}^T(t)]^T\cdot[F(q^{-1})\tilde{\phi}^T(t)] \tag{A3.19}$$

$$R = E\tilde{\phi}(t)\tilde{\phi}^T(t) \tag{A3.20}$$

have the same nullspace.

Proof. Let r be a constant nθ-vector and put $p(t) \triangleq \tilde{\phi}^T(t)r$. Then we can write

$$r^TR_Fr = E[F(q^{-1})p(t)]^T\cdot[F(q^{-1})p(t)] = \mathrm{tr}E[F(q^{-1})p(t)]\cdot[F(q^{-1})p(t)]^T$$

$$= \frac{1}{2\pi}\int_{-\pi}^{\pi} \mathrm{tr}\; F(e^{i\omega})\phi_{pp}(e^{i\omega})F^T(e^{-i\omega})d\omega$$

where $\phi_{pp}(\cdot)$ is the spectral density matrix of p(t). Hence we have the following series of equivalent relations

$$r \in N(R_F) \Leftrightarrow r^TR_Fr = 0 \Leftrightarrow F(e^{i\omega})\phi_{pp}(e^{i\omega})F^T(e^{-i\omega}) \equiv 0 \Leftrightarrow \phi_{pp}(e^{i\omega}) \equiv 0$$

$$\Leftrightarrow Ep^T(t)p(t) = 0 \Leftrightarrow r^T[E\tilde{\phi}(t)\tilde{\phi}^T(t)]r = 0 \Leftrightarrow r \in N(R)$$

The second equivalence is true since tr(A) = 0, with A hermitian and positive semi-definite, implies A = 0. The third follows from the assumption that $\det\{F(e^{i\omega})\} \neq 0$ for any $\omega \in [-\pi,\pi]$. The remaining ones are obvious. The proof is thus finished.

■

Corollary. Under the given conditions, R_F is nonsingular if and only if R is nonsingular. ■

Lemma A3.8. Consider the following matrix of dimension $m_1|m_2$

$$P(D,A,u,m_1,m_2) = E\begin{bmatrix}\frac{1}{D(q^{-1})}u(t-1)\\ \vdots \\ \frac{1}{D(q^{-1})}u(t-m_1)\end{bmatrix}\cdot\left[\frac{1}{A(q^{-1})}u(t-1)\ldots\frac{1}{A(q^{-1})}u(t-m_2)\right] \tag{A3.21}$$

Assume that

i) $m_1 \geq m_2$

ii) A(z) is a polynomial of degree na with all zeros strictly outside the unit circle.

iii) 1/D(z) is a rational function with all poles and zeros strictly outside the unit circle.

Then rank $P(D,A,u,m_1,m_2) = m_2$ if either of the following two conditions is satisfied.

I) D(z)/A(z) is a strictly positive real function and u(t) is persistently exciting or order m_2.

II) The process u(t) is an ARMA(nf,ng) process, i.e.

$$F(q^{-1})u(t) = G(q^{-1})w(t) \tag{A3.22}$$

where w(t) is white noise and the polynomials

$$F(q^{-1}) = 1+f_1q^{-1}+\ldots+f_{nf}q^{-nf}$$

$$G(q^{-1}) = 1+g_1q^{-1}+\ldots+g_{ng}q^{-ng}$$

are relatively prime and have all zeros outside the unit circle. Furthermore, the following condition

$$\max(m_2+ng,\ na+nf) \le m_1 \tag{A3.23}$$

is assumed to hold.

Proof. Consider first part I). It is sufficient to show that the upper square part of $P(D,A,u,m_1,m_2)$ is nonsingular. However, with

$$\varphi(t) = \frac{1}{D(q^{-1})}\,[u(t-1)\ldots\ u(t-m_2)]^T$$

$$H(q^{-1}) = \frac{D(q^{-1})}{A(q^{-1})}$$

the assertion follows from Lemma A3.3.

Consider then part II). Introduce

$$h = [h_1 \ldots\ h_{m_2}] \qquad H(z) = \sum_{i=1}^{m_2} h_i z^i$$

Then, using also the description (A3.22) the equation

$$P(D,A,u,m_1,m_2)h = 0$$

can be written equivalently as

$$E\Big[\frac{G(q^{-1})}{D(q^{-1})F(q^{-1})}\,w(t-j)\Big]\Big[\frac{H(q^{-1})G(q^{-1})}{A(q^{-1})F(q^{-1})}\,w(t)\Big] = 0, \qquad j = 1,\ldots,\ m_1$$

or

$$\frac{1}{2\pi i}\oint z^j\,\frac{G(z)}{D(z)F(z)}\,\frac{H(z^{-1})G(z^{-1})}{A(z^{-1})F(z^{-1})}\,\frac{dz}{z} = 0, \qquad j = 1,\ldots,\ m_1 \tag{A3.24}$$

where the integration path is the unit circle.

The basic tool in determining the solutions h of (A3.24) is Lemma A2.2. To facilitate the application of that lemma, (A3.24) is rewritten as

$$\frac{1}{2\pi i}\oint z^{j-1}\,\frac{G(z)\cdot z^{m_2}H(z^{-1})\cdot z^{ng}G(z^{-1})z^{\max(k,0)}}{D(z)F(z)z^{na}A(z^{-1})\cdot z^{nf}F(z^{-1})z^{\max(-k,0)}}\,dz = 0, \qquad j = 1,\dots,m_1$$

where

$$k = na+nf-m_2-ng$$

It can easily be seen that the number of poles inside the unit circle is $na+nf+\max(-k,0)$. The number of zeros inside the unit circle, neglecting the factor z^{j-1}, is less than or equal to $m_2-1+ng+\max(k,0)$. It then follows from Lemma A2.2 that the only solution will be $h = 0$ (implying rank $P(D,A,u,m_1,m_2) = m_2$) provided

$$m_2-1+ng+\max(k,0) < na+nf+\max(-k,0) \le m_1$$

which can easily be transformed into (A3.23). ∎

Lemma A3.9. Let $z_1(t)$ and $z_2(t)$ be two vector-valued stationary stochastic processes having the same dimensions. Assume that the matrices $Ez_i(t)z_j^T(t)$, $i,j = 1,2$ are nonsingular. Then

$$[Ez_2(t)z_1^T(t)]^{-1}[Ez_2(t)z_2^T(t)][Ez_1(t)z_2^T(t)]^{-1} \ge [Ez_1(t)z_1^T(t)]^{-1} \qquad \text{(A3.25)}$$

where the equality holds if and only if

$$z_1(t) = Mz_2(t) \qquad \text{w p 1} \qquad \text{(A3.26)}$$

M being a constant and nonsingular matrix.

Proof. We clearly have

$$E\begin{bmatrix} z_1(t) \\ z_2(t) \end{bmatrix} [z_1^T(t)\; z_2^T(t)] \triangleq \begin{bmatrix} Z_{11} & Z_{12} \\ Z_{12}^T & Z_{22} \end{bmatrix} \ge 0$$

which gives

$$Z_{11}-Z_{12}Z_{22}^{-1}Z_{12}^T \ge 0$$

and

$$(Z_{12}^T)^{-1}Z_{22}Z_{12}^{-1} \geq Z_{11}^{-1}$$

since the inverses are assumed to exist. Thus (A3.25) is proved.

It is trivial to see that (A3.26) implies equality in (A3.25). Consider now the converse situation, i.e. assume that equality is valid in (A3.25). We then have

$$Z_{11}-Z_{12}Z_{22}^{-1}Z_{12}^T = 0$$

Take $M = Z_{12}Z_{22}^{-1}$. We then get

$$E[z_1(t)-Mz_2(t)][z_1(t)-Mz_2(t)]^T = Z_{11}-MZ_{12}^T-Z_{12}M^T+MZ_{22}M^T$$

$$= Z_{12}Z_{22}^{-1}Z_{12}^T-Z_{12}Z_{22}^{-1}Z_{12}^T-Z_{12}Z_{22}^{-1}Z_{12}^T+Z_{12}Z_{22}^{-1}Z_{12}^T = 0$$

which shows (A3.26) ■

We end this appendix with a lemma proving the nonsingularity of a certain matrix of trigonometric functions.

Lemma A3.10. Consider the matrix

$$H = \begin{bmatrix} 1 & 1 & \dots & 1 \\ \cos\omega_1 & \cos\omega_2 & & \cos\omega_n \\ \cos 2\omega_1 & \cos 2\omega_2 & & \cos 2\omega_n \\ \vdots & \vdots & & \vdots \\ \cos(n-1)\omega_1 & \cos(n-1)\omega_2 & \dots & \cos(n-1)\omega_n \end{bmatrix} \tag{A3.27}$$

Assume that

$$0 \leq \omega_1 < \omega_2 < \dots < \omega_n \leq \pi \tag{A3.28}$$

Then H is nonsingular.

Proof. The proof is similar to that commonly used when evaluating the determinant of a Vandermonde matrix, see e.g. Bellman (1970). We first note that

$$\cos k\omega = \mathrm{Re}[e^{ik\omega}] = \mathrm{Re}[(\cos\omega + i\sin\omega)^k]$$

$$= \sum_{j=0}^{k} \binom{k}{j} \mathrm{Re}[(\cos\omega)^{k-j}(i)^j(\sin\omega)^j]$$

$$= \sum_{\nu=0}^{[k/2]} \binom{k}{2\nu}(\cos\omega)^{k-2\nu}(-1)^\nu(1-\cos^2\omega)^\nu$$

$$\triangleq C_{k,k}(\cos\omega)^k + C_{k,k-2}(\cos\omega)^{k-2} + \ldots \tag{A3.29}$$

In (A3.29) [x] denotes the largest integer less or equal to x. The exact expression of the coefficients $C_{k,i}$ is not important for us, but we note that

$$C_{k,k} = \binom{k}{0} + \binom{k}{2} + \ldots = \sum_{\nu=0}^{[k/2]} \binom{k}{2\nu}$$

always is positive. According to (A3.29) we can regard det(H) as a polynomial of degree n-1 in $\cos\omega_1$. Moreover, H would certainly be singular if $\omega_1 = \omega_i$, any $i > 1$. Thus, det(H) must have the following form

$$\det(H) = \prod_{i=2}^{n} (\cos\omega_1 - \cos\omega_i) f(\cos\omega_2, \ldots, \cos\omega_n)$$

where $f(\cdot, \ldots, \cdot)$ is a polynomial in $\cos\omega_2, \ldots, \cos\omega_n$. Proceeding similarly for $\cos\omega_2$ etc we find that

$$\det(H) = h \prod_{i=1}^{n-1} \prod_{j=i+1}^{n} (\cos\omega_i - \cos\omega_j) \tag{A3.30}$$

where h is a constant. It remains to show that h is nonzero. In the right hand side of (A3.30) h is the coefficient of $(\cos\omega_1)^{n-1}(\cos\omega_2)^{n-2}\ldots(\cos\omega_{n-1})$. It then follows from (A3.27), (A3.29) that

$$h = (-1)^{[n/2]} \prod_{k=1}^{n-1} C_{k,k}$$

which clearly is nonzero. Hence H is nonsingular. ∎

Appendix 4

SOME PROBABILITY RESULTS

In this appendix some convergence results for stochastic variables are given. We say that a stochastic vector x_n *converges in distribution* to $F(x)$ (as n tends to infinity) if the distribution function of x_n converges (pointwise) to $F(x)$. If in particular, $F(x)$ describes a gaussian distribution then we say that x_n is *asymptotically gaussian distributed*.

The main result given in this appendix is a variant of the central limit theorem due to Ljung (1977 a).

Lemma A4.1. Consider

$$X_N = \frac{1}{\sqrt{N}} \sum_{t=1}^{N} z(t) \tag{A4.1}$$

where $z(t)$ is a (vector-valued) zero mean stationary process given by

$$z(t) = \phi(t)v(t) \tag{A4.2}$$

In (A4.2) $\phi(t)$ is a matrix and $v(t)$ a vector. The entries of $\phi(t)$ and $v(t)$ are stationary, possibly correlated, ARMA processes with zero means and underlying white noise sequences with finite fourth order moments. The elements of $\phi(t)$ may also contain a bounded deterministic term.

Then X_N is asymptotically gaussian distributed

$$X_N \xrightarrow{\text{dist}} N(0,P) \tag{A4.3}$$

where, assuming the limit exists

$$P = \lim_{N\to\infty} EX_N X_N^T \tag{A4.4}$$

Proof. See Ljung (1977 a). ∎

The following result on convergence in distribution will often be useful as a complement to the above lemma.

Lemma A4.2. Let $\{x_n\}$ be a sequence of random variables that converges in distribution to $F(x)$. Let $\{A_n\}$ be a sequence of random matrices that converges in probability to A, and $\{b_n\}$ a sequence of random vectors that converges in probability to b. Define

$$y_n = A_n x_n + b_n \tag{A4.5}$$

Then y_n converges in distribution to $F(A^{-1}(y-b))$.

Proof. The lemma is a trivial extension to the multivariable case of the scalar result, given e.g. by Chung (1968), p 85 and Cramér (1946), p 245. ∎

We then specialize the lemma to the case when $F(x)$ corresponds to a gaussian distribution.

Corollary. Assume that x_n is asymptotically gaussian distributed $N(0,P)$. Then y_n as given by (A4.5) converges in distribution to $N(b, APA^T)$.

Proof. The limiting distribution function of y_n is given by ($m = \dim x_n = \dim y_n$)

$$G(y) = \int_{-\infty}^{x} \frac{1}{(2\pi)^{m/2}(\det(P))^{1/2}} e^{-\frac{1}{2}x'^T P^{-1} x'} dx' \Big|_{x=A^{-1}(y-b)}$$

$$= \frac{1}{\det A} \int_{-\infty}^{y} \frac{1}{(2\pi)^{m/2}(\det(P))^{1/2}} e^{-\frac{1}{2}[A^{-1}(y'-b)]^T P^{-1}[A^{-1}(y'-b)]} dy'$$

$$= \int_{-\infty}^{y} \frac{1}{(2\pi)^{m/2}[\det(APA^T)]^{1/2}} e^{-\frac{1}{2}(y'-b)^T(APA^T)^{-1}(y'-b)} dy'$$

Thus we can conclude that $G(y)$ is the distribution function of $N(b, APA^T)$. ∎

In (A4.5) the new sequence y_n is an affine transformation of x_n. For *rational* functions we have another result that concerns convergence in probability. It is given in the following and is often referred to as Slutzky's lemma.

Lemma A4.3. Let $\{x_n\}$ be a sequence of random vectors that converges in probability to a constant x. Let $f(\cdot)$ be a rational function and suppose that f(x) is finite. Then $f(x_n)$ converges in probability to f(x).

Proof. See Cramér (1946), p 254. ∎

Appendix 5

THE LEVINSON-DURBIN ALGORITHM

In this appendix we will give a brief account of the Levinson-Durbin algorithm (LDA). Some of the most useful properties of the algorithm will also be reviewed.

Let $r_0, r_1, r_2, \ldots$ be a sequence of real numbers. Consider the following system of n+1 linear equations with $a_{n,1}, \ldots, a_{n,n}$ and σ_n as unknowns.

$$\begin{bmatrix} r_0 & r_1 & r_2 & \cdots & r_{n-1} \\ r_1 & r_0 & r_1 & \cdots & r_{n-2} \\ \vdots & \vdots & & & \vdots \\ r_{n-1} & r_{n-2} & & \cdots & r_0 \end{bmatrix} \begin{bmatrix} a_{n,1} \\ a_{n,2} \\ \vdots \\ a_{n,n} \end{bmatrix} = - \begin{bmatrix} r_1 \\ r_2 \\ \vdots \\ r_n \end{bmatrix} \tag{A5.1}$$

$$\sigma_n = r_0 + a_{n,1} r_1 + \ldots + a_{n,n} r_n$$

The eqs. (A5.1) are the well-known Yule-Walker equations which appear when fitting an autoregression of order n to the data covariances. This will be discussed in some detail a little later. Here, we keep the discussion deliberately vague since there is no need for such an interpretation of (A5.1) when deriving the LDA.

The LDA, see Levinson (1949), Durbin (1960), exploits the Toeplitz structure of the system matrix in (A5.1) to solve (A5.1) in $O(n^2)$ operations instead of $O(n^3)$ operations that would be needed if the particular structure of (A5.1) would be ignored. Note first that the equations of (A5.1) can be rewritten in the following more compact form

$$[1 \;\; a_{n,1} \cdots a_{n,n}] \begin{bmatrix} r_0 & r_1 & \cdots & r_n \\ r_1 & r_0 & \cdots & r_{n-1} \\ \vdots & \vdots & & \vdots \\ r_n & r_{n-1} & \cdots & r_o \end{bmatrix} = [\sigma_n \; 0 \ldots 0] \tag{A5.2}$$

The basic idea of the LDA is to solve (A5.2) iteratively in n.

Introduce

$$\alpha_n = r_{n+1}+a_{n,1}r_n+\ldots+a_{n,n}r_1 \tag{A5.3}$$

and note that by combining (A5.2) and (A5.3) we get

$$[1 \;\; a_{n,1}\cdots a_{n,n} \;\; 0]\begin{bmatrix} r_0 & r_1 \cdots r_{n+1} \\ r_1 & r_0 \cdots r_n \\ \vdots & \vdots \quad\;\; \vdots \\ r_{n+1} & r_n \cdots r_0 \end{bmatrix} = [\sigma_n \;\; 0 \ldots 0 \;\; \alpha_n] \tag{A5.4}$$

Now, because of the Toeplitz structure of the matrix on the left hand side of (A5.4), we can rewrite (A5.4) as

$$[0 \;\; a_{n,n}\cdots a_{n,1} \;\; 1]\begin{bmatrix} r_0 & r_1 \cdots r_{n+1} \\ r_1 & r_0 \cdots r_n \\ \vdots & \vdots \quad\;\; \vdots \\ r_{n+1} & r_n \cdots r_0 \end{bmatrix} = [\alpha_n \;\; 0 \ldots 0 \;\; \sigma_n] \tag{A5.5a}$$

In order to obtain an updated version of (A5.2) add $-\alpha_n/\sigma_n$ times (A5.5a) to (A5.4). We then get

$$[1 \;\; a_{n,1} - \frac{\alpha_n}{\sigma_n} a_{n,n} \;\ldots\; a_{n,n} - \frac{\alpha_n}{\sigma_n} a_{n,1} \;\; -\frac{\alpha_n}{\sigma_n}]$$

$$\cdot\begin{bmatrix} r_0 & r_1 \cdots r_{n+1} \\ r_1 & r_0 \cdots r_n \\ \vdots & \vdots \quad\;\; \vdots \\ r_{n+1} & r_n \cdots r_0 \end{bmatrix} = [\sigma_n - \frac{\alpha_n^2}{\sigma_n} \;\; 0 \ldots 0] \tag{A5.5b}$$

Compare now (A5.2) and (A5.5b). These equations have the same structure, but the dimension in (A5.5b) is one unit larger. If we assume that $a_{n,1},\ldots, a_{n,n}, \sigma_n$ are known we can from (A5.5b) directly find $a_{n+1,1},\ldots, a_{n+1,n+1}, \sigma_{n+1}$. Since we can easily find the solution of (A5.1) for n = 1 we end up with following recursion called the Levinson-Durbin algorithm (here abbreviated LDA).

$$\begin{cases} a_{k+1,k+1} = -(r_{k+1}+a_{k,1}r_k+\ldots+a_{k,k}r_1)/\sigma_k \\ a_{k+1,i} = a_{k,i}+a_{k+1,k+1}a_{k,k+1-i} \qquad i = 1,\ldots, k \\ \sigma_{k+1} = \sigma_k(1-a_{k+1,k+1}^2) \\ a_{1,1} = -r_1/r_0 \qquad \sigma_1 = r_0-r_1^2/r_0 \\ k = 1, 2,\ldots, n-1 \end{cases} \tag{A5.6}$$

We now proceed to discuss some interesting properties of the LDA. For convenience of the subsequent analysis let us introduce

$$R_n = \begin{bmatrix} r_0 & r_1 & \ldots & r_n \\ r_1 & r_0 & \ldots & r_{n-1} \\ \vdots & \vdots & & \vdots \\ r_n & r_{n-1} & \ldots & r_0 \end{bmatrix} \tag{A5.7}$$

$$A_n(z) = 1 + a_{n,1}z+\ldots+a_{n,n}z^n \tag{A5.8}$$

Lemma A5.1. The following two statements are equivalent:

i) R_n is positive definite $(R_n > 0)$

ii) $A_n(z) \neq 0$ for $|z| \leq 1$, and $r_0 > 0$

Proof. The implication i) → ii) can be shown in many ways. We will present here the proof of Stoica and Söderström (1981 c), see also Mullis and Roberts (1976), Anderson and Moore (1979) for similar proofs. Other types of proofs are given by Tretter (1972), Markel and Gray (1973) and Söderström and Stoica (1981 b).

Assume that $R_n > 0$, and let

$$\bar{r}_n = [r_1 \ldots r_n]^T$$

Consider the following discrete-time system of order n

$$x(t+1) = Ax(t) \tag{A5.9a}$$

where

$$A = \begin{bmatrix} -a_{n,1} & 1 & & \bigcirc \\ \vdots & & \ddots & \\ -a_{n,n-1} & \bigcirc & & 1 \\ -a_{n,n} & 0 & \dots & 0 \end{bmatrix} = \left[\begin{array}{c|c} R_{n-1}^{-1}\bar{r}_n & I_{n-1} \\ \hline & \bigcirc \end{array}\right] \qquad \text{(A5.9b)}$$

Since A is a companion matrix associated with $A_n(z)$ it follows that ii) holds if and only if (A5.9) is asymptotically stable. Consider the following (possible) Lyapunov function

$$V(x) = x^T R_{n-1} x$$

Due to the assumption $R_n > 0$, $V(x)$ is positive definite. Furthermore, we have

$$V(x(t)) - V(x(t+1)) = x^T(t)[R_{n-1} - A^T R_{n-1} A]x(t)$$

with

$$R_{n-1} - A^T R_{n-1} A = \left[\begin{array}{c|c} r_0 & \bar{r}_{n-1}^T \\ \hline \bar{r}_{n-1} & R_{n-2} \end{array}\right] - \left[\begin{array}{c|c} \bar{r}_n^T R_{n-1}^{-1} & \\ \hline I_{n-1} & \bigcirc \end{array}\right] R_{n-1} \left[\begin{array}{c|c} R_{n-1}^{-1}\bar{r}_n & I_{n-1} \\ \hline & \bigcirc \end{array}\right]$$

$$= \begin{bmatrix} r_0 - \bar{r}_n^T R_{n-1}^{-1} \bar{r}_n & \bigcirc \\ \bigcirc & \bigcirc \end{bmatrix} = \begin{bmatrix} \sigma_n & \bigcirc \\ \bigcirc & \bigcirc \end{bmatrix}$$

which is nonnegative definite. The stability of (A5.9) is thus proved. It remains to show that there is no solution $x(t) \not\equiv 0$ such that $V(x(t))$ is constant. Let $x_i(t)$ be the i-th component of $x(t)$. Notice that $V(x(t))$ = constant implies $x_1(t) \equiv 0$. However, the system equation (A5.9) gives then directly $x_j(t+1) = x_{j+1}(t)$ for $j = 1, \dots, n-1$. It can therefore be concluded that $x_j(t) \equiv 0$ for all j. This observation completes the proof of the implication i) $\Rightarrow$ ii).

Assume now that $A_n(z)$ fulfils ii). Consider the following autoregression of order n

$$A_n(q^{-1})u(t) = \varepsilon(t) \qquad \text{(A5.10)}$$

where $\varepsilon(t)$ is white noise ($E\varepsilon^2(t) > 0$). Let

$$r_k^u \triangleq Eu(t)u(t-k) \qquad k = 0, 1, 2, \ldots$$

and introduce R_n^u and $\bar{r}_n^u$ similarly to R_n and $\bar{r}_n$, respectively. Since (A5.10) is asymptotically stable (and hence invertible), the covariances r_k^u, $k = 0,\ldots, n$ fulfil the well-known Yule-Walker equations:

$$\begin{bmatrix} r_o^u & \cdots & r_{n-1}^u \\ \vdots & & \vdots \\ r_{n-1}^u & \cdots & r_o^u \end{bmatrix}\begin{bmatrix} a_{n,1} \\ \vdots \\ a_{n,n} \end{bmatrix} = -\begin{bmatrix} r_1^u \\ \vdots \\ r_n^u \end{bmatrix} \Leftrightarrow R_{n-1}^u \cdot R_{n-1}^{-1}\,\bar{r}_n = \bar{r}_n^u \tag{A5.11}$$

$$r_o^u + [a_{n,1} \ldots a_{n,n}]\bar{r}_n^u = E\varepsilon^2(t).$$

Some straightforward calculations show that (A5.11) implies

$$R_{n-1}^u = A^T R_{n-1}^u A + \begin{bmatrix} E\varepsilon^2(t) & \bigcirc \\ \bigcirc & \bigcirc \end{bmatrix} \tag{A5.12}$$

with A being defined in (A5.9b). Since A is a stability matrix, the Lyapunov equation (A5.12) has a unique solution R_{n-1}^u for every $E\varepsilon^2(t)$, see, e.g., Kailath (1980). It then follows that (A5.11) has also a unique solution w.r.t. r_k^u, $k = 0,\ldots, n$. It is clearly given by

$$r_k^u = \alpha r_k \qquad k = 0,\ldots, n \tag{A5.13a}$$

with α a positive constant (recall that $r_o > 0$) given by

$$\alpha = \frac{E\varepsilon^2(t)}{r_o - \bar{r}_n^T R_{n-1}^{-1}\bar{r}_n} = \frac{E\varepsilon^2(t)}{\sigma_n} \tag{A5.13b}$$

Equation (A5.13) shows that the sequence r_k, $k = 0,\ldots, n$ must be positive definite. The proof is thus finished. ■

Using the calculations made above when proving the implication ii) ⇒ i) we can easily show the following result of independent interest (see also Mullis and Roberts (1976)).

Lemma A5.2. Assume that the sequence $r_o, r_1,\ldots, r_n$ is positive definite (or, in other words, that it is a covariance sequence). Consider the following autoregression of order n (cf (A5.10))

$$u(t)+a_{n,1}u(t-1)+\dots+a_{n,n}u(t-n) = \varepsilon(t) \qquad E\varepsilon(t)\varepsilon(s) = \sigma_n\delta_{t,s} \tag{A5.14}$$

where $\{a_{n,i}\}$ and σ_n are given by (A5.1). Denote by r_k^u the covariance of u(t) at lag k. Then

$$r_k^u = r_k \qquad k = 0,\dots, n \tag{A5.15}$$

that is, the autoregression (A5.14) matches the given covariance sequence $\{r_0,\dots, r_n\}$ exactly.

Proof. The result follows trivially from (A5.13) ∎

An important point which appears in relation to the above lemmas concerns the ability to test whether a given sequence $\{r_0,\dots, r_n\}$ is positive definite or not. Introduce

$$\phi_k = -a_{k,k} \qquad k = 1, 2,\dots \tag{A5.16}$$

where $\{a_{k,k}\}$ are given by (A5.6). If $\{r_0,\dots, r_n\}$ is indeed a sequence of covariances, then $\{\phi_1,\dots, \phi_n\}$ are called the ***partial autocorrelations*** and have a nice statistical interpretation, see, e.g. Box and Jenkins (1976), Ramsey (1974). The following property of $\{\phi_k\}_{k=1}^n$ provides a simple mean to check if the corresponding $\{r_k\}_{k=0}^n$ form a positive definite sequence or not.

Lemma A5.3. The following statements are equivalent:

i) $|\phi_k| < 1 \qquad k = 1,\dots, n$, and $r_0 > 0$

ii) $R_n > 0$

iii) $A_n(z) \neq 0$ for $|z| \leq 1$, and $r_0 > 0$

Proof. Consider the readily verified identity (see, e.g., (A5.2))

$$\left[\begin{array}{c|c} 1 & a_{n,1} \cdots a_{n,n} \\ \hline 0 & I_n \end{array}\right] R_n \left[\begin{array}{c|c} 1 & 0 \\ \hline a_{n,1} & \\ \vdots & I_n \\ a_{n,n} & \end{array}\right] = \left[\begin{array}{c|c} \sigma_n & 0 \\ \hline 0 & R_{n-1} \end{array}\right] \tag{A5.17}$$

which shows that

$$\det(R_n) = \sigma_n \det(R_{n-1})$$

Noting that (A5.6) implies

$$\sigma_k = r_o \prod_{j=1}^{k} (1-\phi_j^2) \qquad k = 1,\ldots$$

we conclude that for any $k \in [1,n]$

$$\det(R_k) = r_o \prod_{i=1}^{k} \sigma_i = r_o^{k+1} \prod_{i=1}^{k} (1-\phi_i^2)^{k-i+1} \tag{A5.18}$$

The equivalence i) $\Leftrightarrow$ ii) follows easily from (A5.18). The other equivalences follow then from Lemma A5.1. However, in order to provide additional insigths into the properties of the LDA we shall prove that i) $\Leftrightarrow$ iii) without recourse to Lemma A5.1. The basic tool will be Rouché's theorem (see e.g. texts on analytic functions like Titchmarsh (1932), or Pearson (1974), p. 255. Markel and Gray (1973) describe the use of this theorem in the present context): If $f(z)$ and $g(z)$ are analytic inside and on a closed contour C, and $|g(z)| < |f(z)|$ for z on C, then $f(z)$ and $f(z)+g(z)$ have the same number of zeros inside C.

It follows from (A5.6), (A5.16) that

$$A_{k+1}(z) = A_k(z)-\phi_{k+1}A_k^*(z) \tag{A5.19}$$

where

$$A_k^*(z) = z^{k+1}A_k(z^{-1})$$

Note that both $A_k(z)$ and $A_k^*(z)$ are analytic for all z.

Assume first that i) holds. Then we have

$$|A_k(z)| = |A_k^*(z)| > |\phi_{k+1}||A_k^*(z)| \qquad \text{for } |z| = 1 \tag{A5.20}$$

Let $A_k(z)$ have all its zeros strictly outside the unit circle. It then follows from (A5.19), (A5.20) and Rouché's theorem that

$$A_{k+1}(z) = 0 \Rightarrow |z| > 1 \tag{A5.21}$$

Since $A_1(z) \neq 0$ for $|z| \leq 1$ $(|a_{1,1}| = |\phi_1| < 1)$ it follows that (A5.21) holds for $k = 0, 1, 2, \ldots$ and thus the implication i) $\Rightarrow$ iii) is proved.

Consider next the implication iii) $\Rightarrow$ i). We have shown above that (A5.21) holds if

$$A_k(z) \neq 0 \quad \text{for} \quad |z| \leq 1, \text{ and } |\phi_{k+1}| < 1 \tag{A5.22}$$

We will now prove that the conditions (A5.22) are also necessary for (A5.21) to be true. Since

$$A_{k+1}(z) = 1 + \ldots + (-\phi_{k+1})z^{k+1} \tag{A5.23}$$

it readily follows that (A5.21) implies $|\phi_{k+1}| < 1$. Assume then that $|\phi_{k+1}| < 1$, but let $A_k(z)$ have, say ℓ zeros within the unit circle. It then follows from the discussion around (A5.20) that $A_{k+1}(z)$ has precisely ℓ zeros inside the unit circle. If $A_k(z)$ has a zero, z^* let us say, on the unit circle then it follows from (A5.19) that z^* will also be a zero of $A_{k+1}(z)$. We can thus conclude that (A5.21) implies (A5.22). The implication iii) $\Rightarrow$ i) follows then easily, since we trivially have $|\phi_1| < 1 \Leftrightarrow A_1(z) \neq 0$, $|z| \leq 1$. ∎

It is worth noting that Lemma A5.3 can be slightly extended to include also cases with $|\phi_k| = 1$, for some k. Note that since the LDA should be stopped when a $|\phi_k| = 1$ is obtained, we can assume without restricting the generality that $|\phi_n| = 1$. The following result dealing with this case can be seen as an addendum to Lemma A5.3.

Lemma A5.4. Assume that $|\phi_k| < 1$, $k = 1, \ldots, n-1$, and $r_0 > 0$. Then the following statements are equivalent:

i) $|\phi_n| = 1$

ii) $\det(R_n) = 0$

iii) $A_n(z) = 0 \Rightarrow |z| = 1$

Proof. The equivalence i) ⇔ ii) is a trivial consequence of (A5.18). The implication iii) ⇒ i) is immediate, cf. (A5.23). It remains to show that i) ⇒ iii). Let ϕ_n^ε be obtained from ϕ_n, $|\phi_n| = 1$, by a small perturbation, and let (cf (A5.19)),

$$A_n^\varepsilon(z) = A_{n-1}(z)-\phi_n^\varepsilon A_{n-1}^*(z)$$

For $|\phi_n^\varepsilon| < 1$, according to Lemma A5.3, $A_n^\varepsilon(z)$ will have all its zeros outside the unit circle. For $|\phi_n^\varepsilon| > 1$ we have

$$|A_{n-1}(z)| < |\phi_n^\varepsilon||A_{n-1}^*(z)| \quad \text{for } |z| = 1$$

and thus, according to Rouché's theorem, $A_n^\varepsilon(z)$ and $A_{n-1}^*(z)$ have the same number of zeros inside the unit circle. However, this means that $A_n^\varepsilon(z)$ will now have all its zeros within the unit circle. Since the zeros of a polynomial are continuous functions of its coefficients we can conclude that iii) must hold. ■

Appendix 6

SOME TERMINOLOGY ASSOCIATED WITH RATIONAL MATRICES

In this appendix we will review some terminology associated with rational (in particular, polynomial) matrices. The discussion will necessarily be brief. For more details as well as for proofs of the assertions see, e.g., Kailath (1980).

A rational (transfer function) matrix $G(q^{-1})$ (i.e., a matrix whose entries are rational functions of the unit delay operator q^{-1}) is said to be *proper* if $G(0) < \infty$ and *strictly proper* if $G(0) = 0$.

A left matrix fraction description (MFD) of G(z),

$$G(z) = A^{-1}(z)B(z) \tag{A6.1}$$

(with z a complex variable replacing q^{-1} above), is said to be *irreducible* if the polynomial matrices A(z) and B(z) are left coprime. A(z) and B(z) are *left coprime* polynomials if they only have *unimodular* matrices (i.e. matrices whose determinant is a nonzero constant, independent of z) as common left divisors. Any rational matrix G(z) has infinitely many irreducible MFDs. They all are related by unimodular left transformations.

Let ν_i be the degree of the i:th row of a (nonsingular) polynomial matrix A(z) (the degree of a polynomial vector is by definition the highest degree of all the entries of the vector). Let A_r be the matrix whose i:th row comprises the coefficients of z^{ν_i} in the i:th row of A(z). Then, A(z) is said to be *row-proper* if A_r is nonsingular. Any (nonsingular) polynomial matrix can be made row-proper by multiplication from the left with an appropriate unimodular matrix.

Consider an irreducible left MFD of G(z), (A6.1), with the polynomial A(z) being row-proper. Then the row degrees of A(z), ν_i, are called the *observability indices* or the *left Kronecker invariants* of G(z). Furthermore

$$\nu = \max_i\{\nu_i\}$$

is the *observability index*, and

$$\delta = \sum_i \nu_i$$

is the ***minimal degree/order*** of G(z). It can be shown that δ is also equal to the determinantal degree of the denominator of any irreducible (left of right) MFD of G(z). Also, ν_i are invariant to unimodular left transformations. Finally, note that for no other left MFD of G(z), the row degrees of A(z) can be smaller than ν_i.

A similar terminology is associated with right MFDs. In that case we can speak, with obvious changes, about right-coprime MFDs, column-proper polynomial matrices, ***controllability indices*** or right Kronecker invariants etc.

REFERENCES

M.S. Ahmed (1982)
Structure determination and parameter estimation of multivariable systems by instrumental variables. *Proc. 6th IFAC Symposium on Identification and System Parameter Estimation*, Washington D.C.

H. Akaike (1971)
Information theory and an extension of the maximum likelihood principle. *2nd International Symposium on Information Theory*, Tsahkadsor, Armenian SSR. Also published in *Supplement to Problems of Control and Information Theory*, pp 267-281, 1973.

H. Akaike (1981)
Modern development of statistical methods. In P. Eykhoff, ed.: *Trends and Progress in System Identification*. Pergamon Press, Oxford.

R.E. Andeen and P.P. Shipley (1963)
Digital adaptive flight control system for aerospace vehicles. *AIAA Journal*, vol 1, pp 1105-1110.

B.D.O. Anderson and E.I. Jury (1976)
Generalized Bezoutian and Sylvester matrices in multivariable linear control. *IEEE Transactions on Automatic Control*, vol AC-21, pp 551-556.

B.D.O. Anderson and J.B. Moore (1979)
Optimal Filtering. Prentice Hall, Englewood Cliffs.

K.J. Åström (1968)
Lectures on the identification problem - the least squares method. Report 6806, Division of Automatic Control, Lund Institute of Technology, Lund, Sweden.

K.J. Åström (1970)
Introduction to Stochastic Control Theory. Academic Press, New York.

K.J. Åström and P. Eykhoff (1971)
System identification - a survey. *Automatica*, vol 7, pp 123-162.

K.J. Åström and T. Söderström (1974)
Uniqueness of the maximum likelihood estimates of the parameters of an ARMA model. *IEEE Transactions on Automatic Control*, vol AC-19, pp 769-773.

G. Banon and J. Aguilar-Martin (1972)
Estimation linéaire recurrente de paramètres des processus dynamiques soumis a des perturbations aleatoires. *Revue du Cethedec*, vol 9, pp 39-86.

I. Barrett-Lenard and J.R. Blair (1981)
Estimation of coefficients for multiple input system models without employing common denominator structure. *International Journal of Control*, vol 33, pp 123-136.

B. Bauer and H. Unbehauen (1978)
On-line identification of a load-dependent heat exchanger in closed loop using a modified instrumental variable method. *Proc. IFAC 7th World Congress*, Helsinki.

R. Bellman (1970)
Introduction to Matrix Analysis. Second edition. McGraw-Hill, New York.

A.J.W. van den Boom (1982)
System identification. On the variety and coherence in parameter and order estimation methods. Doctoral Dissertation, Department of Electrical Engineering, Eindhoven University of Technology, the Netherlands.

G.E.P. Box and G.M. Jenkins (1976)
Time Series Analysis - Forecasting and Control. Holden Day, San Francisco.

J.A. Cadzow (1980)
High performance spectral estimation - a new ARMA model. *IEEE Transactions on Acoustics, Speech and Signal Processing*, vol ASSP-28, pp 524-529.

J.A. Cadzow (1982)
Spectral estimation: an overdetermined rational model equation approach. *Proceedings of IEEE*, vol 70, pp 907-939.

P.E. Caines (1976a)
Prediction error identification methods for stationary stochastic processes. *IEEE Transactions on Automatic Control*, vol AC-21, pp 500-505.

P.E. Caines (1976b)
On the asymptotic normality of instrumental variable and least squares estimators. *IEEE Transactions on Automatic Control*, vol AC-21, pp 598-600.

P.E. Caines and L. Ljung (1976)
Prediction error estimators: Asymptotic normality and accuracy. *Proc. IEEE Conference on Decision and Control*, Clearwater Beach, Florida.

Y.T. Chan (1973)
Parameter estimation of linear multivariable plants using the instrumental variable method. *Proc. 3rd IFAC Symposium on Identification and System Parameter Estimation*, the Hague.

Y.T. Chan and R.P. Langford (1982)
Spectral estimation via the high-order Yule-Walker equations. *IEEE Transactions on Acoustics, Speech and Signal Processing*, vol ASSP-30, pp 689-698.

K.L. Chung (1968)
A Course in Probability Theory. Harcourt, Brace and World, New York.

H. Cramér (1946)
Mathematical Methods of Statistics. Princeton University Press, Princeton.

P.J. Dhrymes, L.R. Klein and K. Steiglitz (1970)
Estimation of distributed lags. *International Economic Review*, vol 11, pp 235-250.

B.W. Dickinson, T. Kailath and M. Morf (1974)
Canonical matrix fraction and state-space descriptions for deterministic and stochastic linear systems. *IEEE Transactions on Automatic Control*, vol AC-19, pp 656-667.

K. Diekmann and H. Unbehauen (1979)
Recursive identification of multi-input, multi-output systems. *Proc. 5th IFAC Symposium on Identification and System Parameter Estimation*, Darmstadt.

J. Durbin (1960)
The fitting of time series models. *Rev. Inst. Int. Statist.*, vol 28, pp 233-244.

H. El-Sherief and N.K. Sinha (1979a)
Online identification of linear discrete-time multivariable systems. *Proceedings of IEE*, vol 126, pp 1321-1325.

H. El-Sherief and N.K. Sinha (1979b)
Choice of models for the identification of linear multivariable discrete-time systems. *Proceedings of IEE*, vol 126, pp 1326-1330.

R.F. Engle (1980)
Exact maximum likelihood methods for dynamic regressions and band spectrum regressions. *International Economic Review*, vol 21, pp 391-406.

P. Eykhoff (1974)
System Identification: Parameter and State Estimation. Wiley, London.

P. Eykhoff (1980)
System identification: approach to a coherent picture through template functions. *Electronic Letters*, vol 16, pp 502-504.

P. Eykhoff, ed. (1981)
Trends and Progress in System Identification. Pergamon Press, Oxford.

B.M. Finigan (1976)
Generation of an asymptotically optimal instrumental variable estimation process using a time-varying auxiliary linear system model. *Proc. 4th IFAC Symposium on Identification and System Parameter Estimation*, Tbilisi, USSR.

B.M. Finigan and I.H. Rowe (1973)
On the identification of linear discrete time system model using the instrumental variable method. *Proc. 3rd IFAC Symposium on Identification and System Parameter Estimation*, the Hague.

B.M. Finigan and I.H. Rowe (1974)
Strongly consistent parameter estimation by the introduction of strong instrumental variables. *IEEE Transactions on Automatic Control*, vol AC-19, pp 825-830.

R. Fletcher (1971)
Fortran subroutines for minimization by quasi-Newton methods. Report AERE-R7125, Harwell.

B. Friedlander (1982a)
Instrumental variable methods for ARMA spectral estimation. *Proc. IEEE International Conference on Acoustics, Speech and Signal Processing*, Paris. Also published in *IEEE Transactions on Acoustics, Speech and Signal Processing* (to appear).

B. Friedlander (1982b)
The overdetermined recursive instrumental variable method. *IEEE Transactions on Automatic Control* (to appear).

K.F. Gauss (1809)
Teoria Motus Corporum Coelestium in Sectionibus Conicus Solem Ambientieum. Reprinted Translation: "*Theory of the motion of the heavenly bodies moving about the sun in conic sections*", Dover, New York.

A. Gauthier and I.D. Landau (1978)
On the recursive identification of multi-input, multi-output systems. *Automatica*, vol 14, pp 609-614.

S. Gentil (1972)
Etude comparative de diverses methodes statistiques d'identification des systèmes dynamiques. Thése de 3ème cycle. Université de Grenoble, Grenoble.

S. Gentil, J.P. Sandraz and C. Foulard (1973)
Different methods for dynamic identification of an experimental paper machine. *Proc. 3rd IFAC Symposium on Identification and System Parameter Estimation*, the Hague.

W. Gersch (1970)
Estimation of the autoregressive parameters of a mixed autoregressive moving average time series. *IEEE Transactions on Automatic Control*, vol AC-15, pp 583-588.

G.C. Goodwin and R.L. Payne (1973)
Design and characterization of optimal test signals for linear single input - single output parameter estimation. *Proc. 3rd IFAC Symposium on Identification and System Parameter Estimation*, the Hague.

G.C. Goodwin and R.L. Payne (1977)
Dynamic System Identification: Experiment Design and Data Analysis. Academic Press, New York.

R. Guidorzi (1975)
Canonical structures in the identification of multivariable systems. *Automatica*, vol 11, pp 361-374.

I. Gustavsson, L. Ljung and T. Söderström (1977)
Identification of processes in closed loop - identifiability and accuracy aspects. *Automatica*, vol 13, pp 59-75.

I. Gustavsson, L. Ljung and T. Söderström (1981)
Choice and effect of different feedback configurations. In P. Eykhoff, ed.: *Trends and Progress in System Identification*. Pergamon Press, Oxford.

R. Haber (1979)
Parametric identification of nonlinear dynamic systems based on correlation functions. *Proc. 5th IFAC Symposium on Identification and System Parameter Estimation*, Darmstadt.

E.J. Hannan (1969)
The identification of vector mixed autoregressive-moving average systems. *Biometrika*, vol 56, pp 223-225.

E.J. Hannan (1971)
The identification problem for multiple equation systems with moving average errors. *Econometrica*, vol 39, pp 751-766.

E.J. Hannan (1975)
The estimation of ARMA models. *The Annals of Statistics*, vol 3, pp 975-981.

E.J. Hannan (1976)
The identification and parameterization of ARMAX and state space forms. *Econometrica*, vol 44, pp 713-723.

J.A. Hausman (1975)
An instrumental variable approach to full information estimators for linear and certain nonlinear econometric models. *Econometrica*, vol 43, pp 727-738.

D.F. Hendry (1976)
The structure of simultaneous equations estimators. *Journal of Econometrics*, vol 4, pp 51-88.

R. Isermann and U. Baur (1974)
Two-step process identification with correlation analysis and least squares parameter estimation. *Trans. ASME*, ser. G., 96, pp 425-432.

R. Isermann, U. Baur, W. Bamberger, P. Kneppo and H. Siebert (1974)
Comparison of six on-line identification and parameter estimation methods. *Automatica*, vol 10, pp 81-103.

A.J. Jakeman (1979)
Multivariable instrumental variable estimators - the choice between alternatives. *Proc. 5th IFAC Symposium on Identification and System Parameter Estimation*, Darmstadt.

A.J. Jakeman, L.P. Steele and P.C. Young (1980)
Instrumental variable algorithms for multiple input systems described by multiple transfer functions. *IEEE Transactions on Systems, Man and Cybernetics*, vol SMC-10, pp 593-602.

A.J. Jakeman and P.C. Young (1979)
Refined instrumental variable methods of recursive time-series analysis. Part II: Multivariable systems. *International Journal of Control*, vol 29, pp 621-644.

A.J. Jakeman and P.C. Young (1981)
On the decoupling of system and noise model parameter estimation in time-series analysis. *International Journal of Control*, vol 34, pp 423-431.

G.M. Jenkins and D.G. Watts (1969)
Spectral Analysis and Its Applications. Holden-Day, San Francisco.

R.H. Jones (1980)
Maximum likelihood fitting of ARMA models to time series with missing observations. *Technometrics*, vol 22, pp 389-395.

P. Jorion and R. Hanus (1980)
Comparative aspects of the on-line instrumental variables method. *Journal A*, vol 21, pp 183-188.

P. Joseph, J. Lewis and J. Tou (1961)
Plant identification in the presence of disturbances and application to digital adaptive systems. *Trans. AIEE on Applications and Industry*, vol 80, pp 18-24.

E.I. Jury (1974)
Inners and Stability of Dynamic Systems. Wiley, New York.

T. Kailath (1980)
Linear Systems. Prentice Hall, Englewood Cliffs.

R.L. Kashyap and R.E. Nasburg (1974)
Parameter estimation in multivariable stochastic difference equations. *IEEE Transactions on Automatic Control*, vol AC-19, pp 784-797.

R.L. Kashyap and A.R. Rao (1976)
Dynamic Stochastic Models from Empirical Data. Academic Press, New York.

I. Kaufman (1969)
The inversion of the Vandermonde matrix and the transformation to the Jordan canonical form. *IEEE Transactions on Automatic Control*, vol AC-14, pp 774-777.

M.G. Kendall and S. Stuart (1961)
The Advanced Theory of Statistics, vol II. Griffin, London.

R.M.C. de Keyser (1979)
On the relationship between some well-known recursive identification methods - the recursive tally principle. *Proc. 5th IFAC Symposium on Identification and System Parameter Estimation*, Darmstadt.

C.E. Kim and J.T. Cain (1982)
An instrumental variable method for stochastic subsystem identification. *Proc. 6th IFAC Symposium on Identification and System Parameter Estimation*, Washington, D.C.

I.D. Landau (1974)
A survey of model reference adaptive techniques - theory and applications. *Automatica*, vol 10, pp 353-379.

I.D. Landau (1979)
Adaptive Control. The Model Reference Approach. Marcel Dekker, New York.

N. Levinson (1949)
The Wiener RMS (root mean square) error criterion in filter design and prediction. In N. Wiener: *Extrapolation, Interpolation and Smoothing of Stationary Time Series*. Wiley, New York.

L. Ljung (1971)
Characterization of the concept of "persistently exciting" in the frequency domain. Report 7119, Division of Automatic Control, Lund Institute of Technology, Lund, Sweden.

L. Ljung (1976)
On the consistency of prediction error identification methods. In R.K. Mehra and D.G. Lainiotis, eds.: *System Identification - Advances and Case Studies*. Academic Press, New York.

L. Ljung (1977a)
Some limit results for functionals of stochastic processes. Report LiTH-ISY-I-0167, Department of Electrical Engineering, Linköping University, Linköping, Sweden.

L. Ljung (1977b)
On positive real functions and the convergence of some recursive schemes. *IEEE Transactions on Automatic Control*, vol AC-22, pp 539-551.

L. Ljung (1978)
Convergence analysis of parametric identification methods. *IEEE Transactions on Automatic Control*, vol AC-23, pp 770-783.

L. Ljung and T. Söderström (1983)
Theory and Practice of Recursive Identification. MIT Press, Cambridge, Mass.

L. Ljung and E. Trulsson (1981)
Adaptive control based on explicit criterion minimization. *Proc. 8th IFAC Congress*, Kyoto.

E. Lyttkens (1974)
The iterative instrumental variables method and the full information maximum likelihood method for estimating interdependent systems. *J. Multiv. Analysis*, vol 4, pp 283-307.

S. Makridakis and S.C. Wheelwright (1978)
Forecasting Methods and Applications. John Wiley & Sons, Santa Barbara.

J.D. Markel and A.H. Gray, Jr. (1973)
On autocorrelation equations as applied to speech analysis. *IEEE Transactions on Audio and Electroacoustics*, vol AU-21, pp 69-79.

D.Q. Mayne (1967)
A method for estimating discrete time transfer functions. In *Advances in Computer Control*, Second UKAC Control Convention, The University of Bristol.

R.K. Mehra (1971)
On-line identification of dynamic systems with application to Kalman filtering. *IEEE Transactions on Automatic Control*, vol AC-16, pp 12-21.

R.K. Mehra (1974)
Optimal input signals for parameter estimation in dynamic systems - A survey and new results. *IEEE Transactions on Automatic Control*, vol AC-19, pp 753-768.

R.K. Mehra (1976)
Synthesis of optimal inputs for multiinput-multioutput systems with process noise. In R.K. Mehra and D.G. Lainiotis, eds.: *System Identification - Advances and Case Studies*. Academic Press, New York.

R.K. Mehra (1981)
Choice of input signals. In P. Eykhoff, ed.: *Trends and Progress in System Identification*. Pergamon Press, Oxford.

R.K. Mehra and D.G. Lainiotis, eds. (1976)
System Identification - Advances and Case Studies. Academic Press, New York.

P.E. Modén (1981)
Offset-free stochastic control with self-tuner applications. Doctoral dissertation, Acta Universitatis Upsaliensis, 613, Uppsala, Sweden.

P.E. Modén and T. Nybrant (1980)
Adaptive control of rotary drum driers. *Proc. 6th IFAC/IFIP International Conference on Computer Applications to Process Control*, Düsseldorf.

C.T. Mullis and R.A. Roberts (1976)
The use of second-order information in the approximation of discrete-time linear systems. *IEEE Transactions on Acoustics, Speech and Signal Processing*, vol ASSP-24, pp 226-238.

R.N. Pandya (1972)
A class of bootstrap estimators for linear system identification. *International Journal of Control*, vol 15, pp 1091-1104.

R.N. Pandya (1974)
A class of bootstrap estimators and their relationship to the generalized two stage least squares estimators. *IEEE Transactions on Automatic Control*, vol AC-19, pp 831-835.

R.N. Pandya and B. Pagurek (1973)
Two stage least squares estimators and their recursive approximation. *Proc. 3rd IFAC Symposium on Identification and System Parameter Estimation*, the Hague.

C.E. Pearson, ed. (1974)
Handbook of Applied Mathematics. Van Nostrand Reinhold, New York.

V. Peterka and A. Halousková (1970)
Tally estimate of Aström model for stochastic systems. *Proc. 2nd IFAC Symposium on Identification and System Parameter Estimation*, Prague.

V. Peterka and K. Šmuk (1969)
On-line estimation of dynamic parameters from input-output data. *4th IFAC Congress*, Warsaw.

J.A. de la Puente and P. Albertos (1979)
Closed loop identification of a steam superheater. *Proc. 5th IFAC Symposium on Identification and System Parameter Estimation*, Darmstadt.

F.L. Ramsey (1974)
Characterization of the partial autocorrelation function. *The Annals of Statistics*, vol 2, pp 1296-1301.

O. Reiersøl (1941)
Confluence analysis by means of lag moments and other methods of confluence analysis. *Econometrica*, vol 9, pp 1-23.

I.H. Rowe (1970)
A bootstrap method for the statistical estimation of model parameters. *International Journal of Control*, vol 12, pp 721-738.

S. Sagara and K. Wada (1977)
An instrumental variable estimator with lagged output as instrument. *Elect. Engng. Jap.*, vol 97, pp 122-128.

S. Sagara, H. Gotanda and K. Wada (1982)
Dimensionally recursive order determination of linear discrete time system. *International Journal of Control*, vol 35, pp 637-651.

N.K. Sinha and A. Sen (1975)
Critical evaluation of on-line identification methods. *Proc. IEE*, vol 122, pp 1153-1158.

S. Sinha and P.E. Caines (1977)
On the use of shift register sequences as instrumental variables for the recursive identification of multivariable linear systems. *International Journal of Systems Sciences*, vol 4, pp 131-138.

T. Söderström (1974)
Convergence of identification methods based on the instrumental variable approach. *Automatica*, vol 10, pp 685-688.

T. Söderström (1975)
Ergodicity results for sample covariances. *Problems of Control and Information Theory*, vol 4, pp 131-138.

T. Söderström, L. Ljung and I. Gustavsson (1974a)
A comparative study of recursive identification methods. Report 7427, Department of Automatic Control, Lund Institute of Technology, Lund, Sweden.

T. Söderström, L. Ljung and I. Gustavsson (1974b)
On the accuracy of identification and the design of identification experiments. Report 7428, Department of Automatic Control, Lund Institute of Technology, Lund, Sweden.

T. Söderström, L. Ljung and I. Gustavsson (1978)
A theoretical analysis of recursive identification methods. *Automatica*, vol 14, pp 231-244.

T. Söderström and P. Stoica (1978)
Comparison of some instrumental variable methods - consistency and accuracy aspects. Report UPTEC 7888R, Institute of Technology, Uppsala University, Uppsala, Sweden.

T. Söderström and P. Stoica (1979a)
Optimal instrumental variable estimation. Report UPTEC 7971R, Institute of Technology, Uppsala University, Uppsala, Sweden.

T. Söderström and P. Stoica (1979b)
Comparison of some instrumental variable methods - consistency and accuracy aspects. *Proc. 5th IFAC Symposium on Identification and System Parameter Estimation*, Darmstadt.

T. Söderström and P. Stoica (1980)
On criterion selection in prediction error identification of least squares models. *Bul. Inst. Politehnic Bucuresti*, vol 42, pp 63-68.

T. Söderström and P. Stoica (1981a)
Comparison of some instrumental variable methods - consistency and accuracy aspects. *Automatica*, vol 17, pp 101-115.

T. Söderström and P. Stoica (1981b)
New and simple proofs for stability of linear prediction models. *Bul. Inst. Politehnic Bucuresti*, vol 43, pp 95-100.

V. Solo (1980)
Some aspects of recursive parameter estimation. *International Journal of Control*, vol 32, pp 395-410.

G.W. Stewart (1973)
Introduction to Matrix Computations. Academic Press, New York.

P. Stoica (1981a)
On a procedure for structural identification. *International Journal of Control*, vol 33, pp 1177-1181.

P. Stoica (1981b)
On multivariate persistently exciting signals. *Bul. Inst. Politehnic Bucuresti*, vol 43, pp 59-64.

P. Stoica (1982a)
On the generalized Yule-Walker equations and testing the order of multivariable time series. *International Journal of Control* (to appear).

P. Stoica (1982b)
On full matrix fraction descriptions. *Bul. Inst. Politehnic Bucuresti*, vol 44, (to appear).

P. Stoica, J. Holst and T. Söderström (1981)
Eigenvalue location of certain matrices arising in convergence analysis problems. Research report 26-81, IMSOR, The Technical University of Denmark, Lyngby, Denmark. A shorter version appeared in *Automatica*, vol 18, pp 487-489, 1982.

P. Stoica and T. Söderström (1979)
Consistency properties of an identification method using the instrumental variable principle. *Revue Roumaine des Sciences Techniques, Serie Electrotechnique et Energetique*, vol 24, pp 289-293.

P. Stoica and T. Söderström (1981a)
Asymptotic behaviour of some bootstrap estimators. *International Journal of Control*, vol 33, pp 433-454.

P. Stoica and T. Söderström (1981b)
On the "min-max optimal" instrumental variable method. Report UPTEC 8102R, Institute of Technology, Uppsala University, Uppsala, Sweden.

P. Stoica and T. Söderström (1981c)
Stability properties of autoregressive models fitted to multivariable time series. Report UPTEC 8155R, Institute of Technology, Uppsala University, Uppsala, Sweden.

P. Stoica and T. Söderström (1981d)
Instrumental variable methods for identification of linear and certain nonlinear systems. Report UPTEC 8168R, Institute of Technology, Uppsala University, Uppsala, Sweden.

P. Stoica and T. Söderström (1982a)
Instrumental variable methods for identification of Hammerstein systems. *International Journal of Control*, vol 35, pp 459-476.

P. Stoica and T. Söderström (1982b)
Comments on the Wong and Polak minimax approach to accuracy optimization of instrumental variable methods. *IEEE Transactions on Automatic Control*, vol AC-27, pp 1138-1139.

P. Stoica and T. Söderström (1982c)
Identification of multivariable systems using instrumental variable methods. *Proc. 6th IFAC Symposium on Identification and System Parameter Estimation*, Washington D.C.

P. Stoica and T. Söderström (1982d)
Optimal instrumental-variable methods for the identification of multivariable linear systems. *Automatica*, vol 19, (1983, to appear).

P. Stoica and T. Söderström (1982e)
On the parsimony principle. *International Journal of Control*, vol 36, pp 409-418.

P. Stoica and T. Söderström (1982f)
A useful input parameterization for optimal experiment design. *IEEE Transactions on Automatic Control*, vol AC-27, pp 986-989.

P. Stoica and T. Söderström (1983)
Optimal instrumental variable estimation and approximate implementation. *IEEE Transactions on Automatic Control*, vol AC-28, (to appear).

P. Stoica, T. Söderström, A. Ahlén and G. Solbrand (1983)
On the convergence of pseudolinear regression algorithms. In preparation.

A. Stoica and P. Stoica (1976)
Linear system identification using the tally principle. In *Revue Roumaine des Sciences Techniques, Serie Electrotechnique et Energetique*, vol 21, pp 435-443.

E.C. Titchmarsh (1932)
The Theory of Functions. Oxford University Press, London.

S.A. Tretter (1972)
The all-pole model is theoretically stable. *IEEE Transactions on Audio and Electroacoustics*, vol AU-20, p 316.

S.G. Tzafestas (1970)
Some computer-aided estimators in stochastic control systems identification. *International Journal of Control*, vol 12, pp 385-399.

R.K. Ward (1977)
Notes on the instrumental variable method. *IEEE Transactions on Automatic Control*, vol AC-22, pp 482-484.

P.E. Wellstead (1978)
An instrumental product moment test for model order estimation. *Automatica*, vol 14, pp 89-91.

P.E. Wellstead and R.A. Rojas (1982)
Instrumental product moment model-order testing: extensions and applications. *International Journal of Control*, vol 35, pp 1013-1027.

P. Whitehead and P.C. Young (1979)
Water quality in river systems: Monte-Carlo analysis. *Water Resources Research*, vol 15, pp 451-459.

P. Whitehead, P.C. Young and G. Hornberger (1979)
A systems model of stream flow and water quality in the Bedford-Ouse river-1. Stream flow modelling. *Water Research*, vol 13, pp 1155-1169.

K.Y. Wong and E. Polak (1967)
Identification of linear discrete time systems using the instrumental variable approach. *IEEE Transactions on Automatic Control*, vol AC-12, pp 707-718.

W.M. Wonham (1974)
Linear Multivariable Control - A Geometric Approach. Springer-Verlag, Berlin.

W.R. Wouters (1972)
On-line identification in an unknown stochastic environment. *IEEE Transactions on Systems, Man and Cybernetics*, vol SMC-2, pp 666-668.

P.C. Young (1965a)
On a weighted steepest descent method of process parameter estimation. Report, Cambridge University, Engineering Laboratory, Cambridge, U.K.

P.C. Young (1965b)
Process parameter estimation and self adaptive control. *Proc. IFAC Symposium*, Teddington; appears in P.H. Hammond, ed.: *Theory of Self Adaptive Control Systems*, Plenum Press, New York, 1966.

P.C. Young (1968)
The use of linear regression and related procedures for the identification of dynamic processes. *Proc. 9th IEEE Symposium on Adaptive Processes*, UCLA, Los Angeles, California.

P.C. Young (1970a)
An instrumental variable method for real-time identification of a noisy process. *Automatica*, vol 6, pp 271-287.

P.C. Young (1970b)
An extension to the instrumental variable method for the identification of a noisy dynamic process. Technial note CN/70/1, Control Engineering Group, University of Cambridge, U.K.

P.C. Young (1972)
Comments on "On-line identification of linear dynamic systems with applications to Kalman filtering". *IEEE Transactions on Automatic Control*, vol AC-17, pp 269-270.

P.C. Young (1974)
Recursive approaches to time series analysis. *Bulletin of Inst. of Mathematics and its Applications* (IMA), vol 10, pp 209-224.

P.C. Young (1976)
Some cbservations on instrumental variable methods of time series analysis. *International Journal of Control*, vol 23, pp 593-612.

P.C. Young (1981)
Parameter estimation for continuous-time models - a survey. *Automatica*, vol 17, pp 23-39.

P.C. Young and R. Hastings-James (1970)
Identification and control of discrete linear systems subject to disturbances with rational spectral density. *Proc. 11th IEEE Symposium on Adaptive Processes*, Austin, Texas.

P.C. Young and A.J. Jakeman (1979a)
Refined instrumental variable methods of recursive time-series analysis. Part I: Single input, single output systems. *International Journal of Control*, vol 29, pp 1-30.

P.C. Young and A.J. Jakeman (1979b)
The development of CAPTAIN: a computer aided program for time-series analysis and identification of noisy systems. *Proc. IFAC Symposium on Computer Aided Design of Control Systems*, Zürich.

P.C. Young and A.J. Jakeman (1980)
Refined instrumental variable methods of recursive time-series analysis. Part III: Extensions. *International Journal of Control*, vol 31, pp 741-746.

P.C. Young, A.J. Jakeman and R. McMurtrie (1980)
An instrumental variable method for model order identification. *Automatica*, vol 16, pp 281-294.

P.C. Young, S.H. Shellswell and C.G. Neethling (1971)
A recursive approach to time-series analysis. Report No CUED/B-Control/TR16. Department of Engineering, University of Cambridge, Cambridge, U.K.

P.C. Young and P.G. Whitehead (1977)
A recursive approach to time-series analysis for multivariable systems. *International Journal of Control*, vol 25, pp 457-482.

M.B. Zarrop (1979a)
A Chebyshev system approach to optimal input design. *IEEE Transactions on Automatic Control*, vol AC-24, pp 687-698.

M.B. Zarrop (1979b)
Optimal Experiment Design for Dynamic System Identification. Springer-Verlag, New York.

SUBJECT INDEX

D

E

F

G

H

I

P

R

S

Lecture Notes in Control and Information Sciences

Edited by A. V. Balakrishnan and M. Thoma

Vol. 22: Optimization Techniques
Proceedings of the 9th IFIP Conference on
Optimization Techniques,
Warsaw, September 4–8, 1979
Part 1
Edited by K. Iracki, K. Malanowski, S. Walukiewicz
XVI, 569 pages. 1980

Vol. 23: Optimization Techniques
Proceedings of the 9th IFIP Conference on
Optimization Techniques,
Warsaw, September 4-8, 1979
Part 2
Edited by K. Iracki, K. Malanowski, S. Walukiewicz
XV, 621 pages. 1980

Vol. 24: Methods and Applications
in Adaptive Control
Proceedings of an International Symposium
Bochum, 1980
Edited by H. Unbehauen
VI, 309 pages. 1980

Vol. 25: Stochastic Differential Systems –
Filtering and Control
Proceedings of the IFIP-WG7/1 Working Conference
Vilnius, Lithuania, USSR, Aug. 28 – Sept. 2, 1978
Edited by B. Grigelionis
X, 362 pages. 1980

Vol. 26: D. L. Iglehart, G. S. Shedler
Regenerative Simulation of Response
Times in Networks of Queues
XII, 204 pages. 1980

Vol. 27: D. H. Jacobson, D. H. Martin, M. Pachter, T. Geveci
Extensions of Linear-Quadratic Control Theory
XI, 288 pages. 1980

Vol. 28: Analysis and Optimization of Systems
Proceedings of the Fourth International
Conference on Analysis and Optimization of Systems
Versailles, December 16–19, 1980
Edited by A. Bensoussan and J. L. Lions
XIV, 999 pages. 1980

Vol. 29: M. Vidyasagar,
Input-Output Analysis of Large-Scale
Interconnected Systems –
Decomposition, Well-Posedness and Stability
VI, 221 pages. 1981

Vol. 30: Optimization and Optimal Control
Proceedings of a Conference Held at
Oberwolfach, March 16–22, 1980
Edited by A. Auslender, W. Oettli, and J. Stoer
VIII, 254 pages. 1981

Vol. 31: Berc Rustem
Projection Methods in Constrained
Optimisation and Applications
to Optimal Policy Decisions
XV, 315 pages. 1981

Vol. 32: Tsuyoshi Matsuo,
Realization Theory of
Continuous-Time Dynamical Systems
VI, 329 pages, 1981

Vol. 33: Peter Dransfield
Hydraulic Control Systems –
Design and Analysis of Their Dynamics
VII, 227 pages, 1981

Vol. 34: H.W. Knobloch
Higher Order Necessary Conditions
in Optimal Control Theory
V, 173 pages, 1981

Vol. 35: Global Modelling
Proceedings of the IFIP-WG 7/1 Working
Conference Dubrovnik, Yugoslavia,
Sept. 1–5, 1980
Edited by S. Krčevinac
VIII, 232 pages, 1981

Vol. 36: Stochastic Differential Systems
Proceedings of the 3rd IFIP-WG 7/1
Working Conference
Visegrád, Hungary, Sept. 15–20, 1980
Edited by M. Arató, D. Vermes, A.V. Balakrishnan
VI, 238 pages, 1981

Vol. 37: Rüdiger Schmidt
Advances in Nonlinear
Parameter Optimization
VI, 159 pages, 1982

Vol. 38: System Modeling and Optimization
Proceedings of the 10th IFIP Conference
New York City, USA, Aug. 31 – Sept. 4, 1981
Edited by R.F. Drenick and F. Kozin
XI, 894 pages. 1982

Vol. 39: Feedback Control of
Linear and Nonlinear Systems
Proceedings of the Joint Workshop
on Feedback and Synthesis of
Linear and Nonlinear Systems
Bielefeld/Rom
XIII, 284 pages. 1982

Vol. 40: Y.S. Hung, A.G.J. MacFarlane
Multivariable Feedback:
A Quasi-Classical Approach
X, 182 pages. 1982

Vol. 41: M. Gössel
Nonlinear Time-Discrete Systems –
A General Approach by
Nonlinear Superposition
VIII, 112 pages. 1982

Vol. 42: Advances in Filtering and
Optimal Stochastic Control
Proceedings of the IFIP-WG 7/1
Working Conference
Cocoyoc, Mexico, February 1–6, 1982
VIII, 391 pages. 1982

Lecture Notes in Control and Information Sciences

Edited by A. V. Balakrishnan and M. Thoma

Vol. 43: Stochastic Differential Systems
Proceedings of the 2nd Bad Honnef Conference
of the SFB 72 of the DFG at the University of Bonn
June 28 – July 2, 1982
Edited by M. Kohlmann and N. Christopeit
XII, 377 pages. 1982.

Vol. 44: Analysis and Optimization of Systems
Proceedings of the Fifth International
Conference on Analysis and Optimization of Systems
Versailles. December 14–17, 1982
Edited by A. Bensoussan and J. L. Lions
XV, 987 pages, 1982

Vol. 45: M. Arató
Linear Stochastic Systems
with Constant Coefficients
A Statistical Approach
IX, 309 pages. 1982

Vol. 46: Time-Scale Modeling of Dynamic Networks
with Applications to Power Systems
Edited by J. H. Chow
X, 218 pages. 1982

Vol. 47: P. A. Ioannou, P. V. Kokotovic
Adaptive Systems with Reduced Models
V, 162 pages. 1983

Vol. 48: Yaakov Yavin
Feedback Strategies for Partially
Observable Stochastic Systems
VI, 233 pages, 1983

Vol. 49: Theory and Application of Random Fields
Proceedings of the IFIP-WG 7/1
Working Conference
held under the joint auspices of the
Indian Statistical Institute
Bangalore, India, January 1982
Edited by G. Kallianpur
VI. 290 pages. 1983

Vol. 50: M. Papageorgiou
Applications of Automatic Control Concepts
to Traffic Flow Modeling and Control
IX, 186 pages. 1983

Vol. 51: Z. Nahorski, H.F. Ravn, R.V.V. Vidal
Optimization of Discrete Time Systems
The Upper Boundary Approach
V, 137 pages 1983

Vol. 52: A. L. Dontchev
Perturbations, Approximations and Sensitivity Analysis
of Optimal Control Systems
IV, 158 pages. 1983

Vol. 53: Liu Chen Hui
General Decoupling Theory of Multivariab
Process Control Systems
XI, 474 pages. 1983

Vol. 54: Control Theory for Distributed
Parameter Systems and Applications
Edited by F. Kappel, K. Kunisch,
W. Schappacher
VII, 245 pages. 1983.

Vol. 55: Ganti Prasada Rao
Piecewise Constant Orthogonal Function
and Their Application to Systems and Co
VII, 254 pages. 1983.

Vol. 56: Dines Chandra Saha, Ganti Pras
Identification of Continuous
Dynamical Systems
The Poisson Moment Functional
(PMF) Approach
IX, 158 pages. 1983.

Vol. 57: T. Söderström, P. G. Stoica
Instrumental Variable Methods
for System Identification
VII, 243 pages. 1983.